A HYBRID CONSCIOUSNESS?

RAFAEL PINTOS-LÓPEZ

Copyright © 2025 by Rafael PINTOS-LÓPEZ

All rights reserved.

No part of this book may be reproduced in any form or by any electronic or mechanical means, including information storage and retrieval systems, without written permission from the author, except for the use of brief quotations in a book review.

❦ Created with Vellum

CONTENTS

Preface   v
Introduction   ix

1. How language originated the cognitive process   1
 EAST & WEST   13
2. Boddhidharma   14
 - the rejection of cognition
3. Hezekiah   27
 - writing the soul of a civilisation
 SCIENCE   39
4. Schrödinger   40
 - information and life have different natures
5. Neuroscience   51
 - futile attempts at measuring experience
6. Darwin & Wallace   58
 - divergence over human evolution
 LANGUAGE & CULTURE   73
7. Whorf   74
 – how language influences thought
8. Aristotle, Averroes & Borges   82
 - the story of a mistake
 TIME   91
9. Borges, again   92
 - time and other ideas
10. Schrödinger, again   100
 - time according to scientists
 PHILOSOPHY   107

11. Wittgenstein 108
   - understood by few

12. Schopenhauer 114
   - the original Western visionary

Afterword 119
Acknowledgments 129

PREFACE

"This book will perhaps only be understood by those who have themselves already thought the thoughts which are expressed in it—or similar thoughts."

Tractatus Logico-Philosophicus —

Ludwig Wittgenstein

"Here's to the crazy ones. The misfits. The rebels. The troublemakers. The round pegs in the square holes. The ones who see things differently. They're not fond of rules. And they have no respect for the status quo. You can quote them, disagree with them, glorify or vilify them. About the only thing you can't do is ignore them. Because they change things. They push the human race forward. And while some may see them as the crazy ones, we see genius. Because the people who are crazy

enough to think they can change the world are the ones who do."

Apple Advertisement ——

Steve Jobs

What I propose in the chapters that follow appears based on an 'absurd' premise: that thousands of scientists and philosophers, the world over, are all wrong; that they are all mistaken, and that I am right. It's not like that at all.

What I propose—and I provide evidence to that effect—is that some of the great thinkers of the past are being deliberately and unreasonably overlooked. The general view among scientists and philosophers nowadays is against Cartesian dualism (the separation of body and mind). The view is that consciousness is a physical, individual, phenomenon. They say there is nothing metaphysical about consciousness. They equate consciousness solely with sentience, with experience.

The current trend is wrong on various accounts. I give some examples, but the list is by no means exhaustive. Their premise, for instance, is that human consciousness is located within the individual brain. "Does the self exist outside the brain?" we hear. There is no doubt that the brain has a big role in anything to do with the self and with consciousness.

The self is not a physical entity. It does not exist within the brain. The self includes many phenomena. Interoception and exteroception (which involve the whole body) are part of self-awareness. But the concept of self also includes identity, which is cognitive and has social and cultural implications.

PREFACE

Separating sentience from cognition is almost impossible. It is done, mostly in the East, and under very special circumstances.

Human consciousness is more than sentience. It has included cognition since we became human. I add that cognition and humanity are the result of language. And there is nothing universal or natural about language. Language is cultural and artificial.

INTRODUCTION

"The book will, therefore draw a limit to thinking, or rather—not to thinking, but to the expression of thoughts; for, in order to draw a limit to thinking we should have to be able to think both sides of this limit (we should therefore have to be able to think what cannot be thought).".

Tractatus Logico-Philosophicus —

Ludwig Wittgenstein

"The brute feels and perceives; man, in addition to this 'thinks' and 'knows': both 'will'."

The World as Will and Idea —

Arthur Schopenhauer

INTRODUCTION

This is meant to be reading material for anyone interested in human consciousness. It has a small emphasis on language and culture. It attempts to answer questions that have remained unanswered so far, and it relies on some brilliant minds and how they dealt with transcendental questions. It is about the way religion, philosophy and science—the three main perspectives of the quest for knowledge—have historically approached the issue. It includes seemingly unrelated chapters, some of them weird and some not so weird, but all of them providing evidence towards the same notion that sentience and cognition are the discrete but almost inseparable components of human consciousness. It goes further than that: as the title suggests, I propose some radical ideas that do not constitute a proper theory of human consciousness but that—in a way—I see as a roadmap towards a 'unified field' theory of consciousness.

I submit that we became conscious through a long process that had a feedback loop and input from evolutionary biology, i.e. the neocortex. Towards the end—in all likelihood, between 135,000 and 70,000 years ago—there was an interruption in the process. After that, it became more of a meta-evolutionary phenomenon. Perhaps the beginning of this Introduction is the best place for me to state clearly what I understand by that.

These are my conjectures:

I believe it would be good for philosophy and science to consider human consciousness as a hybrid phenomenon that consists of two integrated but discrete layers or components:

an original animal layer (sentience)—which is natural and biological—, and a subsequent human layer (cognition)—which is human-made and, as such, artificial. They have two different natures. The notion that sentience, by itself, constitutes human consciousness only creates confusion. Deeming the components discrete and defining the boundaries between them is of the utmost importance for a proper understanding of consciousness. Sentience should be considered as an ineffable, fundamental, component of living creatures (language cannot describe the richness of experience, and measuring experience serves no purpose). Sentience does not ask for explanations and does not give them either.

Maybe in the future, a better understanding of quantum physics' particle entanglement would give us a clearer picture of the behaviour of the senses, which would not—by any means— constitute a solution.

No evolutionary, i.e., biological, study of consciousness can reach human consciousness; it is not part of a continuum; the cognitive component adds a different nature to our minds: human consciousness is not totally biological.

Tens of thousands of years ago, there was a point at which evolution gave way to another process. Fully developed syntax was something exclusively human, and also artificial and meta-evolutionary. Language is not natural. Other human phenomena followed language. Basically, *H. sapiens* created language and language created *H. sapiens*. At that moment, our minds had become hybrid, part biological and part metaphysical.

Cognition is only acquired linguistically, through parental and cultural upbringing. It is individually transmitted. This needs to happen every generation, which makes us a highly altricial species. Cognition was and is introduced by language and its nature is as artificial as those of culture and language, otherwise we would be born equipped with full comprehension and speech.

I suspect voluntary imagination, creativity, long-term memory, adventurousness, are exclusively human traits acquired through language and culture, i.e., they did not exist before cognition. They are not evolutionary.

Since thought is produced by language and culture, it is obviously influenced by them (as per Whorfian Linguistic Relativity).

The logical conclusion is that there should be cortical and other specialised centres in the brain—newer than any centre that deals with strictly biological phenomena—where cultural developments are processed (e.g., Broca's and Wernicke's).

Time is a human construct that exists only within cognition, through unlimited voluntary imagination (expectation) and long-term memory (which includes identity and collective perception). The counterpart is that—like in other animal species—human sentience remains limited to present and change.

Perhaps, a *sine qua non* test for full consciousness should be the ability to communicate that one is conscious (in ways to be determined). There are different technological means to establish consciousness in comatose people or people with language disabilities. Any definition of consciousness would benefit from that type of requirement.

INTRODUCTION

The questions have been around for many generations. They involve issues such as life, death, time, imagination, and memory. The answers have never been quite satisfactory. Religion attributed life and consciousness to a deity or deities, i.e., no further explanation possible.

Philosophy has been trying to delve into human consciousness but always encounters the same barrier as science: how can we explain experience? Where is it? Science and philosophy have not found a solution as to how it operates so far. During the past few years—with modern methods, such as new imaging technology—there have been big advances in neuroscience. Why is it that Western science and philosophy still find themselves confronted with an enigma of such proportions?

Human consciousness has been considered a paradoxical phenomenon by many at least since the times of King Hezekiah of Judea, Plato, St Augustine of Hippo, Democritus, and many other, more recent, thinkers. Human beings have senses and, at the same time, think. Sentience and cognition have coexisted within us ever since we became human. They work in tandem and yet they appear incompatible. How does that happen? They complement each other but cognition fails to explain sentience, and sentience is not interested in explanations. Many philosophers and scientists have thought about it. The great Descartes, among the most lucid of men, asserted: *"I think, therefore I am"*. That established a clear causation. Yes, in order to ask yourself something, you have to exist. That does not answer the fundamental questions of life and consciousness. His substance dualism, however, point towards a solution: there is life—which in our case involves sentience, or experience—and then there is thought (not a soul in the religious sense). One is physical and the other one, metaphysical.

Philosophy and science have confused the issue even more by treating human consciousness as just sentience, whereas human consciousness involves two layers and those layers are inextricably united. Without cognition we cannot begin to discuss human consciousness or ask questions about it. The phenomenon is unique.

Having failed to understand Ludwig Wittgenstein's *Tractatus Logico-Philosophicus*, originally published more than a century ago, in 1921, Western scientists and philosophers are still trying to understand experience.

The bottom line is that science—and/or philosophy—may not be able to completely explain human consciousness because they are not equipped to do it, i.e., because they are cognitive.

Is it possible to have some details as to why not? There are some answers.

INTRODUCTION

Perhaps I should begin by analysing where the academic establishment fails to find answers and where our conjectures could provide a way out of the status quo.

I think David Chalmers, a renowned senior philosopher, is representative of how the issue has been approached and why things have ended up in a dead end. Mainstream ideas have gone a certain way, that's all.

In 1995, Chalmers came up with the concept of *"the hard problem of consciousness"*. How can we explain 'subjective' experience? In 1998, Christof Koch, a renowned scientist and author of *"The Quest for Consciousness: a Neurobiological Approach"*, wagered with Chalmers that the problem would be solved within 25 years. Koch lost the bet in 2023. Even finding the correlates of experience in the brain would not have solved the problem. Was there a holistic way to explain consciousness?

As stated above, there are many phenomena that neither science nor philosophy have been able to explain thus far: randomness (entropy) is one of them. Physics has found entropy at the quantum level. Heisenberg's uncertainty principle dictates that the position and speed of a particle, such as a photon or electron, cannot be known simultaneously with perfect accuracy. The same kind of uncertainty occurs at a larger scale. Randomness has eluded a full description or an explanation. But I am digressing. Let's go back to consciousness.

INTRODUCTION

Chalmers summarised his views on a TED talk of 2014, which I believe sheds light on mainstream ideas. Let us analyse some segments of what he says.

He begins the talk: *"<u>Right now you have a movie playing inside your head</u>.*. ... Your movie has smell, and taste, and touch. It has a sense of your body, pain, hunger, orgasms. It has emotions, anger and happiness, <u>it has memories, like scenes from your childhood, playing before you, and it has this constant voiceover narrative in your stream of conscious thinking</u>.*. At the heart of this movie is you, experiencing all this, directly. This movie is your stream of consciousness, <u>the subjective experience of the mind and the world</u>.*. Consciousness is one of the fundamental facts of human existence."*

* *(My underlining)*

Very aptly, he begins by comparing sentience to a movie. That is exactly the way things are in real life. The Hollywood movie, in real life, though, is a cheap 2D version of sentience. It is as if we were witnesses of something that is happening before our very eyes, but that is not actually happening. Sentience works like that, but for real. We witness reality, we witness the universe. And when I say the movie is 'a cheap' version, what I mean is that—as Chalmers very well says—real life is more than a show. It has smell and touch and all the other senses.

Then he describes different ways in which you can feel your body (interoception, if you like), which is all just fine: he is still describing sentience. But then he mentions *"scenes from your childhood..."*, your memory. The moment he says that, he confirms that philosophy and science have the concepts of sentience and cognition all jumbled up. They know they are different phenomena, but they do not take their discreteness into account in the context of human consciousness. One thing is clear: you cannot discuss memory as a component of sentience. Long-term episodic memory, like time and knowledge of human finitude, are products of cognition. In sentience, however, there is no time. Nobody feels hungry in the past. Nobody has an orgasm in the future. Sentience is present. Life happens now.

I would submit that long-term episodic memory only appears in humans, and that it happens only after toddlers acquire language. Maybe I am wrong, but if you think about your own experience, you will notice that your memories (and your identity) begin then, the moment you acquire language, i.e., between two and four years of age.

Later, Chalmers mentions *"a voiceover narrative"*. A movie does not normally need to explain anything. It just shows. Exceptionally, a movie has a voiceover which clarifies what is happening. Maybe the director—somebody like Woody Allen, perhaps— believes the voiceover will add to the drama of a scene, or things need to be clarified. The voiceover within our head, however, is not an explanation. In real life we don't need that explanation either. The voiceover in real life is something totally unrelated. When we do that, we are generally digressing. We are thinking that we have to pick up Sally from school, or leave the shirts for dry cleaning, or maybe we're just stressing about some possible future event. Again, any narrative we may have in our heads takes the form of language, and that is exclusively cognitive and exclusively human. Artificial. The chapter on Boddhidharma explains how that happens.

INTRODUCTION

I would venture that no animal has a voiceover narrative (because they lack language and complex thought), and because they don't need it.

∽

Let us go back to the TED talk.

Chalmers says: *"I want a scientific theory of consciousness that works. And, for a long time, I banged my head against the wall looking for a theory of consciousness in purely physical terms that would work. But I eventually came to the conclusion that it just didn't work for systematic reasons. It's a long story but it's just the core idea of what you get from pure reductionist explanations in physical terms, in brain-based terms, it's stories about the functioning of the system, its structure, its dynamics, the behaviour it produces. Great for solving easy problems. How we behave, how we function, <u>but when it comes to subjective experience, why does all this feel like something from the inside?</u>*. That's something fundamentally new and it's always a further question. So, I think we're in a kind of impasse here. <u>We've got this wonderful great chain of explanation. We're used to it, where physics explains chemistry, chemistry explains biology, biology explains parts of psychology. But consciousness doesn't seem to fit into this picture.</u>*. On the one hand, it's a datum. <u>We're conscious. On the other hand, we don't know how to accommodate it into our scientific view of the world.</u>*. So, I think consciousness right now is a kind of anomaly, one we need to integrate into our view of the world but we don't see how. Faced with an anomaly like this, radical ideas may be needed."*

* *(My underlining)*

Here, he mentions very important elements of the problem: *'a theory... in purely physical terms'*. Language and culture are not purely physical, or 'subjective', phenomena. You cannot reduce them to that. He then asks why *'subjective experience'* feels like something from the inside; again, a fairly Western kind of concept. What he calls the *'subjective experience'* is basically sentience. Sentience does not divide reality into 'subjective' and 'objective'. Sentience just is. Many other species have it. The difference is that, in human beings, it is intertwined with cognition.

In 1911, Bertrand Russell said Wittgenstein could study philosophy under him if he could solve his paradox. Russell's paradox states: *"Let R be the class of all classes that do not contain themselves. If R contains itself, then it is a contradiction. If R does not contain itself, it satisfies the requirement of the class, but it contradicts the assumption that it contains all classes that do not contain themselves. Both cases are contradictory"*.

The paradox was solved several times, but Wittgenstein's solution was radical. The whole concept of classes—he said —is an unwarranted assumption. The problem was the wrong problem from the beginning.

With all due respect, the answer to Chalmers' 'hard problem' is that the question is the wrong question. The question should not be "*... why does all this feel like something from the*

inside?"; it should be "Why am I able to ask the question that this experience comes from the inside?". The answer is that all animals have sentience and that it probably feels like something from the inside to them as well, but they do not feel separate from the rest of reality and, in any case, they do not feel the need to ask or answer any questions. Animals don't ask questions (to themselves or to other animals). Only humans have cognition intertwined with sentience, which allows us to question why we have sentience in the first place.

Some people have told me that the problem with my assumption of a hybrid nature of consciousness is that I mix experience with cognition, when consciousness should only include experience. The answer to that is that I am not the one mixing anything. Philosophers and scientists acknowledge they are both together. The general assumption is that they are components of consciousness: as Chalmers said, childhood memories (which is actually time), an internal narrative (speech), and many other phenomena are included in cognition. The problem is that they do not establish clear boundaries when, in reality, even their natures are different.

'Human' consciousness is different from anything else. Dan Dennett acknowledged this. I would add that the cognitive component of consciousness is not the one that experiences the universe, but—as opposed to what happens to other species—, it is the one that allows us to 'witness' it, i.e., the one that produces our own special version of reality. Cognition is the inquisitive but artificial component of our consciousness, the layer that wants explanations. And that is exactly how Chalmers phrases it. He lists all kinds of scientific explanations. All those we understand—he says—but there is one datum that we cannot understand: *"we're conscious"*. That does not fit into our scientific view of the world—he adds. No, it doesn't—I reply—and it never will, because explanations are neither biological nor sentient. Qualia are ineffable because language is artificial. It cannot deal with them. What is perceiving our consciousness and our witnessing is the component that can understand explanations and artificially explain other phenomena, like behaviour. But—necessarily—it cannot understand that other layer of consciousness and that witnessing because their natures are different: paradoxically, 'subjective' biological phenomena are not subject to explanation because any rational, cognitive, explanation would be artificial. And when I say 'artificial' I mean exactly that, an artifice, something human-made, and something exclusively human. The impossibility, then, lies in the fact that our perception of the world is biological, whereas the linguistic and cultural components of consciousness (and thus of philosophy and neuroscience) are artificial.

Now let us analyse one last time a segment of Chalmers' talk.

"I want to explore two crazy ideas that I think may have some promise. <u>The first crazy idea is that consciousness is fundamental.</u>* ... If you can't explain consciousness in terms of the existing fundamentals: space, time, mass, charge, then it's a matter of logic, you need to expand the list. <u>The natural thing to do is postulate consciousness itself as something fundamental. A fundamental building block of nature.</u>* This doesn't mean you suddenly can't do science with it. This opens up the way for you to do science with it. We then need is to study the

fundamental laws governing consciousness. The laws that connect consciousness to other fundamentals: space, time, mass, physical processes. <u>The second idea is that consciousness might be universal.</u>* .... It's also worth noting that although the idea seems counterintuitive to us, it is much <u>less counterintuitive to people from different cultures where the human mind is seen as much more continuous with nature</u>.*. A deeper motivation comes from the idea that perhaps the most simple and powerful way to find fundamental laws connecting consciousness to physical processing is <u>to link consciousness to information</u>.*. Wherever there is information processing there is consciousness— complex information processing, like in a human, complex consciousness; simple information processing, simple consciousness.

*(My underlining)

Chalmers—basing his proposal on Dan Dennett's theory of consciousness—suggests that consciousness might be considered a fundamental. Here, again, we find that the problem resides in confusing sentience with consciousness as a whole. Cognition is definitely not fundamental, since we can analyse it and explain it. But it is definitely part of human consciousness. When cognition does its own analysing, it becomes meta cognition. It explains itself. Science can analyse and measure the behaviour of the different senses as well. What science and philosophy cannot do—as both science and philosophy are cognitive—is explain sentience. Similarly, science can explain the behaviour of life, but not the experience. Science can measure how a bat flies, or hangs from a branch. It cannot explain what being a bat is like. We cannot even imagine what their senses would feel like. Could we imagine the million colours a hummingbird perceives?

INTRODUCTION

Is consciousness universal? Well, cognition is definitely not universal. It is only human.

The other idea that Chalmers discusses is based on Giulio Tononi's Integrated Information Theory (IIT). A couple of years ago, it was declared one of the most popular theories among neuroscientists and philosophers. Christof Koch is one of the main supporters of Tononi's theory. Even the name of the theory sounds unfortunate. Information and experience have totally different natures. Information is cognitive. The theory's five principles are that any experience exists for itself, it's specific, it's structured, it's unitary and it's definite. The main idea behind it is that we can measure the amount of experience in a system. It appears obvious—at least it appears obvious to me—that, even if it were at all possible—measuring experience would not help in understanding it. We can keep time, and yet we cannot understand how it works. There is no reasoning involved in experience. How can numbers help when numbers are part and parcel of reasoning?

One of the most incredible—and unscientific—possibilities considered by IIT, for instance, is that even inert systems (lifeless objects) could be conscious. The theory suggests that consciousness might be universal. The mere conjecture does not make any sense. Totally unverifiable, or unfalsifiable. A return to panpsychism (!).

As mentioned before, consciousness, as in human consciousness—not just sentience—cannot be universal. But even sentience cannot be universal. Ours is based on the preexistence of very important phenomena, some of which are fundamental: iron> ozone> oxygen> life> sentience> cognition. I would venture that our type of consciousness can only exist on planet Earth.

We know that we have sentience and cognition, that they are different phenomena. What I say here is something much more drastic, something I believe most people do not see or do not want to see. I believe they are not just different. I say both components of our consciousness have different natures and, even though they work at unison within human consciousness, they are totally discrete. Sentience is biological. Cognition is not. It is artificial. Why do I maintain this?

There are many reasons, but there is a very simple explanation which is quite evident but is generally overlooked. We are used to the fact that we are not like other animal species; the thing is we are different, but mainly in one special way. Our animal consciousness accommodates an extra component that is not biological or intuitive. We are not born cognitive, we individually acquire cognition. Like all other species, we are born naturally—i.e., biologically—but we have to be taught cognition; and that happens because it is artificial. In every single one of us. Every generation is the same. Individuals are born without it. Newborn babies cannot speak. They cannot think. And they cannot think because they have no language. Some argue that fetuses have consciousness. That is impossible. They may have a degree of sentience, nothing else. Language is essential for complex thought and without complex thought there is no human consciousness. Language and thought can explain many phenomena, but not sentience.

Some may say that what I affirm here is a truism. It is possible. But no theory of consciousness mentions this. A truism? Yes. But one that keeps on being ignored. Philosophers and neuroscientists are currently looking for an impossibility. Any theory of consciousness has to begin by accepting sentience as a fundamental. The other side of the coin is that language, and the culture that comes with language, are both artificial. They are both a human artifice. A human construct. Concluding that language and thought are historically related leads directly to Language Relativity.

Let us think of identity, for instance. Let us imagine you meet someone at a party. It could be a baptism, a bar mitzvah, or a *vernissage*. At a baptism, the probability would be that the person is Christian, at a bar mitzvah, that he is Jewish, and at a *vernissage,* or exhibition, that he is interested in art. You already have several points from which to analyse who this person—this man—is. Let's choose the exhibition. He approaches you and makes small talk about a painting.

"Hi, how are you?, incidentally, I'm James Horowitz, a friend of Peter's".

The guy is in his fifties, with a beard, has an American accent, probably from the East coast. The surname Horowitz is still an enigma, it could be Jewish or non-Jewish. Not that it really matters, but the surname is originally from the town of Hořovice, in Bohemia, so he is from European stock, as his looks confirm, but the East coast accent tells you he is probably second or third generation American. If you're a woman, you would probably also be checking his appearance as well. He's interested in art, because Peter is a painter. He's probably not in the military, because of the general aspect and beard. Casual looks, so you would say that, politically, his leanings are more liberal than conservative.

Is this natural? Are your attitude and your analysis of this individual something natural? You would probably answer "Yes". We would say that there's nothing natural about it. You have put all kinds of labels on this person. You have classified him and ticked all kinds of boxes. You are trying to analyse his ethnic background, religion and his politics. Of course, there is a reason for that. To establish a relationship or friendship with the person, or even just to do small talk with him, you need information about him. That would be the beginning of "knowing" him.

This is a human thing. And it is not anything new.

A similar encounter between Australian aborigines in the bush would be something like this:

"Do you know Mary Warrlpungi?"

"No."

"What about Joe Nyulu?

"No."

"Maybe you know Betty Ngurraar?"

"Yeah, she's my cousin"

"Ahhh, that girl is my granddaughter; so, you are my grandson too. You have to give me a cigarette."

A relationship has been established. So, you're right, that's what is needed when you meet someone. But then I would still say it's not natural. I would say it's an exclusively human thing. That means there's something artificial about it.

When two individuals of another animal species see each other for the first time, they do not need to know the identity of the other. There is none. They do not need to know family background or history—they have no long-term episodic memory, or time, other than the present. There are no questions. They need no geographic points of reference. This is just another individual. Maybe they need to visually find out about their sex, or maybe they guess it or sniff it. Anything they are going to be interested in is whether this individual is apt for mating, or if he or she wants to establish a territory that may overlap theirs. That is natural. If the individuals are from different species, then, they will have to establish whether this other animal is predator or prey, or whether it is indifferent. That is also natural.

Self-awareness is biological (natural) and present, whereas identity is hybrid and time-related (it includes cognition). You are probably self-aware from the moment you are born (you are sentient and interoceptive). Self-awareness occurs only in the present; you cannot be self-aware in the past or in the future. You can remember or imagine being self-aware, though.

Identity, on the other hand, is hybrid because you acquire your identity through the collective. Human beings need identities in order to function within their social groups.

INTRODUCTION

The time-related aspect of one's identity has to do with the continuity of that identity through the name society gives the individual. Identity is permanent. For the individual and for society. It may vary in the case of gender change, but that is only a socially accepted exception.

Here we cannot help but returning to Heraclitus, the second time the individual crosses the river, the river is the same and the individual is the same, but they are not actually the same. How is that? To some extent— they are both different, pretty much like Theseus' ship: some elements may have gradually changed. The secret of the continuity is the day-by-day imperceptibility of that change.

But identity involves much more than self-awareness. It involves—to a much greater extent—the perception of society rather than that of the individual. Evolution may have introduced 'selfhood'—the term Anil Seth (*Being You*) uses—to keep you alive, but it has not designed 'identity': that is a social creation in which the individual participates. One of the interesting things about the beginning of identity is that it coincides with the beginning of memory. There is a family/group identity—your first name—; and there is a social identity—your surname. They come at different times. And they are both language-related. The first memories we have are as toddlers, when we are beginning to speak and understand. Not before then. Also, we can associate both with the beginning of cognition.

Our species needs lots of non-biological information. I do not have to explain why. It happens. But I need to explain how that operates in terms of our consciousness.

According to Seth:

INTRODUCTION

"An influential tradition, dating back at least as far Descartes in the seventeenth century, held that non-human animals lacked conscious selfhood because they did not have rational minds to guide their behaviour"... "I don't agree. <u>In my view, consciousness has more to do with being alive than with being intelligent.</u>*. <u>We are conscious selves precisely 'because' we are beast machines. I will make the case that experiences of 'being you', or of 'being me', emerge from the way the brain predicts and controls the internal state of the body.</u>*. The essence of <u>self-hood</u>* is neither a rational mind nor an immaterial soul. It is a deeply embodied biological 'process', a process that underpins the simple feeling of being alive that is the basis for all our experiences of self, indeed for any conscious experience at all. 'Being you' is literally about your body".

* (My underlining)

The confusion is evident. It's not that just Seth himself is confused. The trend among neuroscientists and philosophers is that human consciousness is just sentience. But he also equates 'cognition' with 'intelligence'. On top of that, he throws interoception into the mix. Denying that cognition is an inextricable component of human consciousness results in blurry notions of selfhood, identity and self-awareness.

Seth discusses 'selfhood'. I don't understand exactly what he means by that. It's a fuzzy term. I presume he is talking about identity in human beings because he writes of *"being you"* or *"being me"*, and we are both human. If that is the case, he is sadly mistaken. "Being me" is being Rafael Pintos-López. It has nothing to do with an immaterial soul because I assume I do not have one. It has a lot to do with my rational mind, with my cognition. That is the identity I grew up into as a member of human society. I know that for a fact. Also, there is no doubt in my mind that identity involves time and memory. *"Feeling alive"* is something that happens in the here and now. 'Identity' includes the memories of childhood that Chalmers mentions in his TED talk. Non-human animals lack *"conscious selfhood"*, as Seth says, because they do not need it. They are sentient, not conscious. They live in the present. They may have some degree of selfhood in that they feel their own bodies, but I cannot imagine them feeling 'subjective', separate from the rest of nature.

Denying that memory and time have a lot to do with the concept of self in human beings does not make sense. You can divide human self into 'cognitive' and 'sentient' self, or you can call those elements 'epistemic' and 'phenomenal' self. It doesn't matter. Human beings need both to be who they are.

A third example of the trend prevalent among philosophers and scientists, American philosopher Thomas Nagel *(What is it like to be a bat?)* sees the difference between sentience and cognition as a difference clearly linked to a false assumption: that consciousness should be considered within the field of biology and—more specifically—that it should be taken as an evolutionary phenomenon:

"But to explain consciousness, as well as biological complexity, as a consequence of the natural order adds a whole new dimension of difficulty. <u>I am setting aside outright dualism, which would abandon the hope for an integrated explanation. Indeed, substance dualism would imply that biology has no</u>

<u>responsibility at all for the existence of minds.</u>*. What interests me is the alternative hypothesis that biological evolution is responsible for the existence of conscious mental phenomena, but since those phenomena are not physically explainable, the usual view of evolution must be revised. It is not just a physical process."

Agreed, to a point. Substance dualism does not necessarily exclude the biological component of the mind, or of language, for that matter. After he discards substance dualism, he adds a footnote:

"But substance dualism would still leave biology with a huge problem similar to the one we are discussing: namely, <u>why has physical evolution produced organisms of a kind capable of being occupied by and interacting with minds?</u>"*.

* (My underlining)

I believe Nagel is talking about human minds. Cognition—I conjecture—is a "black swan" phenomenon; something totally unexpected, improbable to say the least. Bayesian probabilities cannot account for such phenomena. Cognition is, indeed, a byproduct of language, which, I maintain, is meta-evolutionary. Physical evolution—biology—has not produced organisms that think and interact.

What Nagel is saying, among other things, is that the subdivision of disciplines within the biological sciences is something that needs to be revised. The problem with Nagel's analysis (among many other philosophers) is that he includes both sentience and cognition as a conglomerate within the term 'consciousness'. Sentience is a necessary product of biology, it is a quality that emerges directly from biology.

Sentient animals have evolved using their senses and need them to survive. That is just the way it is: a brute fact of reality. Cognition, however, is not a further complication of sentience. It is a whole new development, a human addition solely related to language and culture. Accounting for cognition on entirely reductive terms is what has taken science to the position it finds itself in right now. It is impossible to advance when the premise of the research is wrong. Philosophy has been asking the wrong question. What is required now is for philosophy to ask the right question and start all over again.

Neuroscience repeats the same ideas. Seth explains:

"It is widely agreed that experience arises from a physical basis, but we have no good explanation of why and how it so arises. <u>Why should physical processing give rise to a rich inner life at all?</u>*. It seems objectively unreasonable that it should, and yet it does."

*(My underlining)

And one wonders why the whole field of neuroscience is going around in circles. Physical processing does not give rise to a rich inner life. Sentience is there. It just is. Only human cognition witnesses nature. It is only human cognition, the meta-evolutionary, metaphysical component of consciousness that produces human inner life.

In this Introduction, I only try to suggest an idea of what "human consciousness" is. This will provide the basis to understanding the rest of the book, which is not about linguistics, or neuroscience, or biology, or philosophy. It's about common sense. Consciousness is a complicated subject, but you will see that many brilliant minds understood, or guessed, that its complication lies in the fact that its nature is hybrid. Without having expressed it in so many words, the famous thinkers I cite in the different chapters of the book, appear to agree with it.

As mentioned above, the approach of this book is neither scientific nor philosophical. I will try to explain why there are things that appear to be ineffable.

HOW LANGUAGE ORIGINATED
THE COGNITIVE PROCESS

"We have found a strange footprint on the shores of the unknown. We have devised profound theories, one after another, to account for its origins. At last, we have succeeded in reconstructing the creature that made the footprint. And lo! It is our own."

Space, Time, and Gravitation –

Sir Arthur Eddington

What are codes? The Dewey Decimal System and the Library of Congress System, for instance, are codes (they literally add order to information by means of a sticker); the Morse Code (a binary information system through space); the German *"Enigma"* Code (during World War II, Germans exchanged information that the

Allies could not understand); Egyptian and Mayan hieroglyphs (which could only be read by a selected few): codes are means of communication between interlocutors who know how to decipher the message.

When our hominin ancestors commenced to talk, they actually invented the first basic code: a phoneme, i.e., sound that could be understood and combined to add meaning. But it was much more; it was something metaphysical and artificial, something that had never existed before, completely human-made: complex meaning. Complex meaning did not exist before humanity. And even now, complex meaning remains exclusively human.

What they did was like attaching a mental sticker to an object or an action, which eventually resulted in infinite combinations of sounds and more: infinite meaning. It became progressively more and more complex. There was a moment, then, that was the culmination of an incredibly lengthy process (tens of thousands of years). That artificial sticker, that meaning or understanding, floated between the interlocutors. It was something they had in common, that united them, something aptly called "communication". But it was also something intangible, something metaphysical that an animal species had created and that, in turn, made its members human (*H. sapiens*): meaning.

A RECENT STUDY of language comprehension conducted by Andrey Vyshedskiy et al, *"Three mechanisms of language are revealed through cluster analysis of individuals with language deficits"* confirms that language comprehension can be divided into clusters of abilities. At a non-individual, macrocosmic, level, those clusters could give us an indication of how language developed historically.

The study classified the clusters, in order of difficulty, as follows: *"The cluster of most-basic abilities, termed 'command-language'comprehension', included knowing the name, responding to 'No' or 'Stop', and following some commands*. The cluster of intermediate abilities, termed 'modifier-language-comprehension', included understanding color and size modifiers, several modifiers in a sentence, size superlatives, and numbers*. The cluster of most-advanced abilites, termed the 'syntactic-language'comprehension' included understanding of spatial prepositions, verb tenses, flexible syntax, possessive pronouns, explanations about people and situations, simple stories, and elaborate fairytales*".*

* *(My underlining)*

In terms of cognition, the levels lead me to believe that comprehension of meaning went from understanding basic commands to increasingly abstract thought. The first level— the most basic and direct one—must have included self-awareness and awareness of the environment (something we share with other species); the second level—increasingly human, I would say—some combination of terms, colours, measurement and numbers (already abstract thought); and the third one—with cognition fully developed—, understanding of space-time, recursive language, and imagination of non-existing entities. Basically, how we went from animal to human in three basic steps. Sounds easy now, but it took many thousands of years.

At an interview with *PsyPost*, Vyshedskiy stated the view that the studies he conducted on language comprehension have confirmed his Whorfian views concerning language: *"For over 50 years, linguists such as Noam Chomsky and Steven Pinker have proposed the existence of a uniquely human language comprehension mechanism, yet its neurological basis remains largely unknown."*. Absolutely. They cannot prove it because it does not exist.

Thus far, there is no proof of a universal "language template" in our brain. What is certain is that individual cultures affect their own languages. If that is the case, it follows that the thought processes of the speakers of any given language are affected by that language.

How does physical matter give rise to metaphysical inner life? The answer is: physical matter doesn't. It is involved, but the reality is much more complicated than that. It is possible to listen to *The Goldberg Variations*, for instance, and cry with emotion. Yet, the origin of the emotion has very little to do with the behaviour of neurones, molecules or synapses.

Meaning was metaphysical from its very origin. Before language there was no meaning. It acquired an existence all of its own that resides neither in the mouth or the mind of the individual that utters the sound nor in the ear or the mind of the individual that deciphers the utterance. It is something we share, something interlocutors have in common that does not belong to either of them, or to any other member of the culture, or species, for that matter. Those ends that utter and decipher meaning are indeed physical, but meaning itself has a life of its own that is historically provided by the collective (by the culture).

The fact that its creation was artificial belies any claims in favour of a purely physical or biological origin of cognition. There is no doubt: the moment a hominin combined terms to add meaning to something and another hominin understood that combined meaning, that was the onset of humanity.

Surely, a more basic kind of non-linguistic 'meaning' existed among other animal species before then: grunts, yells and body language to signify "danger" or "food", as it happens with other social species, like chimpanzees. The difference between those and the first actual phoneme lied in how deliberate the vocal/auditory exchange was and the fact that, potentially, it could be combined and recombined with other sounds to add more meaning to the message. So, perhaps the onset of humanity came about not just because of the creation of the first phoneme, but because of its potential to be combined with other sounds in order to produce syntax (I would venture that morphological variations might have been a latter refinement). In any case, it was a very gradual phenomenon. Basically, 70,000 years ago language evolved from very simple modifiers and acquired its recursive nature, i.e., complex syntax. By recursive we mean that a sentence can be included within another sentence: "Andy said that going ahead with the plan was crazy", where "going ahead with the plan was crazy" is a sentence within a sentence. There is no limit to the possible combinations of sounds and meanings in any given language.

Thought and culture grew together with language, and their growth was exponential. A complex language required memory and that—of course—was fed back into the loop. Human memory expanded, and long-term episodic memories found their place in the neocortex, outside of the hippocampus. That allowed for the existence of time and identity. There were biological consequences, more sophisticated linguistic production and reception organs, and specialist centres in the brain, like Broca's and Wernicke's. The neocortex grew, the brain grew in size, together with the cranium that contained it. Parturition became more complex and painful for the females of the species.

The fact that the neocortex represents roughly ninety percent of the cerebral cortex, and that it is called "neo" (new) means that there is a correlation between its expansion and the use of language, which in my conjecture means the beginning of cognition.

The understanding of time, for instance, is encoded in the parietal cortex. It evolved with the use of simple modifiers until it reached more complex syntax. We can say, with a high degree of certainty, that the process took place 70,000 years ago. Many other human phenomena, like imagination and long-term memory are cognitive in nature.

From all appearances, this creation of meaning was a unique event which, as I said, took tens of thousands of years to reach its climax. And it was also unique in that it had happened only to our species. Using Taleb's terminology, I would call it a "black swan" event, but would go even further and call it the beginning of meta-evolution. Our species would continue to evolve biologically—or not—, but the main source of variations from that moment on would be the new human consciousness.

Let us give the idea some more detail: let us say that the hominins that produced sound with some meaning had remained natural animals, to some extent, like chimpanzees (and we are talking about a very slow progression). But during the creative process, some of them were becoming increasingly human. The moment complex language had become a reality, the very process of creation had converted them into fully operational human beings. It was a mutual phenomenon. Language was being created by *H. sapiens* and *H. sapiens* was being created by language. In the words of Erwin Schrödinger: *"For we ourselves are chisel and statue, conquerors and conquered at the same time — it is a continued 'self-conquering' (Selbstüberwindung)."*.

Yuval Noah Harari calls the culmination of that prolonged process the *"Cognitive Revolution"*. Some neuroscientists still attribute the behavioural change to a *"random genetic mutation"*.

In reality, complex thought appears to have resulted from the development of recursive language. It followed language. Complex thought grew until it developed into a clear distinction between *H. sapiens* and other animal species. In a strange way, something that had expanded as a means of communication—i.e., a social phenomenon—became a process that allowed for an individual, subjective, analysis of reality. Those factors resulted in the possibility of a new witnessing. No creature had witnessed reality before. Not in the same way we understand witnessing.

Human witnessing is what we now understand as human consciousness. No other animal species witnesses as we do. No other individual animal is quite an individual as we are. Individuals of other species are simply iterations of the species. They are part of nature, nothing else. *H. sapiens* is part of nature as well, but not completely. We have an artificial component that allows for a special analysis of what we think 'the rest of nature' is. You can call it 'objective reality', 'otherness', 'that which is not me'. But we are the only species that, having the two components within our consciousness, can ask about experience. No bat, no duck, no tiger, questions its own sentience. They have a degree of consciousness, but that degree of consciousness doesn't include complex cognition. They just experience.

At the beginning of this chapter, I quote Sir Arthur Eddington. When he wrote what he wrote he was referring to quantum physics. The same could be said about the origins of cognition. Our ancestors left their footprint. They created meaning, and life would never be the same again.

LET US ELABORATE ON THIS. Western (Judaeo-Christian) religion in general, decided that human beings were separate from the rest of nature and established a clear boundary between the observer (a human being) and the observed ('objective reality'). Christianity, influenced by Greek philosophy, went one step further. Humans were even more special. The reason was that, with humanity, a new element had emerged that was metaphysical. That other something that had not existed before, that only humans had, was cognition (and, with it, information and communication). The Greeks called it Ψυχή (*psyche*), meaning 'soul'. We now use the same word to mean 'mind'. The first chapter of *East & West* will include the first attempts at describing life and consciousness.

Somehow, Western ideas were correct. That artificial layer of consciousness created individual minds with individual views of reality. But Eastern ideas were also correct: it is also possible for human beings to achieve oneness with nature.

The feedback loop between mind and meaning (communication) resulted in another metaphysical entity: collective consciousness. Individuals could share information with other individuals and all that knowledge became a metaphysical corpus. After long periods of sharing information, humans had common knowledge. That corpus of metaphysical knowledge was shared orally for some time, until writing was created. Nowadays, it is shared by other technological means as well. The study of individual brains will lead to a better understanding of human consciousness only when its metaphysical component is included.

Before cognition, *H. sapiens* used the same system other species had been using for aeons: sentience. Sentience meant that individuals could interact with their environment without setting themselves apart from it. For the time being, let us treat sentience as a fundamental phenomenon.

Let us say that, with language, human beings had devised a codified system that could be added to their biological minds. But those two entities, the code and the biological organs that produced it and received it had totally different natures: one was artificial and the other ones, natural. The biological organs had to evolve and adapt to a new reality.

There probably was some basic thought without language—I would say the germ of thought, 'ideas', which are not quite the same as thoughts, and this is evident in other species; but we believe we can safely state that the most important discrete component of human consciousness—cognition—is a byproduct of social communication, i.e., of complex language. And we can safely state it because there is clear evidence that cognition stagnates in isolated human infants.

There are many examples that tend to confirm this hypothesis: during dreams the sense of identity is muted; there is self-awareness. There may be some language, but there is no logic involved in dreams. They are dynamic images, like watching a movie, but without any rhyme or reason attached to them. Dreams seem to be basically biological with a tiny cognitive input in humans. Writing in dreams becomes a very frustrating activity. We may know that writing exists, but we cannot do it. The characters are not clear and we are not dexterous enough to write them. Maybe we can understand what something written means, but that is as far as it goes.

In the chapter on *Boddhidharma*, I emphasise that cognition is artificial. It is a human creation that has to be taught every generation. We are naturally born as other mammalian species, without cognition. Human individuals continue to be born without cognition as we have done for aeons. Collective rearing and socialisation adds that artificial layer to our consciousness.

We are what biologists would call an altricial species; by that they mean that our young take a long time to grow up and become independent. In a way—this is controversial and we will come back to this point—that could also be expressed by saying that our offspring take a long time to become fully and operationally human.

Cognition makes us what we are: human beings, but cognition has limitations inherent in its artificiality. When I say that it makes us what we are, I imply that we are meta-evolutionary creatures, an unintended result of a unique phenomenon. We humans are a byproduct of the creation of meaning.

From what I have been saying, it seems quite apparent that I consider sentience and cognition as having totally different natures. The only thing cognition (largely aided by language) and experience (totally biological) have in common is that they are both components of human consciousness. The development of a theory of consciousness is hindered by that duality. The artificial layer of our consciousness can be explained. Sentience can only be shown. The biological part of our consciousness is directly experienced by the individual. It is individual and cannot be shared through language.

The fact that it coexists with the human cognitive layer makes it the subject of human inquisitiveness. Other animal species may be curious, but they are not inquisitive. Without cognition, crocodiles cannot be curious about their sentience (sentience is not curious about itself). We can ask: how is it that we experience? Where are the correlates of experience in our brain? Those questions are possible because of the second layer of our consciousness.

EAST & WEST

BODDHIDHARMA

- THE REJECTION OF COGNITION

"How hard, then, and yet how easy it is to understand the truth of Zen! Hard because to understand it is not to understand it; easy because not to understand it is to understand it. A Master declares that even Buddha Sakyamuni and Bodhisatva Maitreya do not understand it, where simple-minded knaves do understand it."

An Introduction to Zen Buddhism -

D.T. Suzuki

"An intrinsic property is traditionally understood as a property that something would have even if it were the only thing in the universe or the only thing in existence. Does that idea even make sense? Not if you think that something is what it is only by virtue of its

belonging to a web of relations. Why not say that relations determine the occupants of the relations, after the fashion of relational quantum mechanics? Or that relations and occupants are mutually interdependent?"

Frank, Gleiser & Thompson - *The Blind Spot*

"The over-all number of minds is just one. I venture to call it indestructible since it has a peculiar timetable, namely mind is always now. There is really no before and after for mind. There is only a now that includes memories and expectations."

Erwin Schrödinger - *What is life?*

This chapter will deal with human consciousness and Zen. An Eastern view of consciousness, if you like.

For millennia, East and West have held opposed—seemingly irreconcilable—views of what human consciousness is. The East appeared to have chosen wisdom, and the West, knowledge. On different occasions throughout history, the East kept on choosing sentience and meditation; the West chose cognition and inquisitiveness. Experience versus thought. Both profoundly valid.

By asserting that human consciousness has two separate natures, i.e., that it is a hybrid phenomenon, I am not actually saying that both, Eastern and Western schools of thought are correct. On the contrary, but they are both needed to understand our mind. Thus far, Western science and philosophy have tended to ignore the East altogether.

Zen is not a religion, nor is it a philosophy. It has no definition. Zen is a return to nature.

I will try to explain how Zen unravels the components of human consciousness. The process began with Buddha in Nepal and took centuries to evolve, until Boddhidharma distilled its essence in China. His pupil, Dōgen Zenji, picked up the baton, and this eventually led to the founding of the Sōtō school of Zen in Japan.

They go where science cannot go. Which means that, sometimes, not understanding is all right. You will be the judge as to how logical or credible the conjectures are.

THE WORLD—GAUTAMA said—is contingent. Change is inevitable. Being born, growing up, getting sick, becoming old and dying are all aspects of time, whereas the senses only accept what happens here and now: the present.

Gautama had evidently guessed the dual nature of human consciousness. Sentience, he discovered, was the key to *nirvana;* utter wisdom came only through direct sensorial experience and nature. Within sentience there was nothing to be explained, no future, nothing to stress about.

Buddha said that the highest state a human can aim for is ecstasy, which is achieved through meditative concentration. *Nirvana* was the only form of salvation. Buddhism rejected the separate existences of consciousness and matter, subject and object, soul and deity. There was one reality. Only a dream that has no dreamer. The dream is surrounded by nothingness.

Eastern schools of thought maintain that everything is connected. And that is true. Much more than people normally understand. At a much more intimate level. That is not science, but it has a scientific explanation. Tom Chi, an astrophysicist and polymath, explains it brilliantly. I recommend his videos. Maybe a Western mind needs an explanation like that. It needs to 'understand' why things are the way they are.

As with the rest of other animals, our hearts beat in order to carry a molecule called haemoglobin through our bloodstream. Every molecule of haemoglobin has a single iron atom (Fe^{II}). So iron is absolutely essential for us to be alive and stay alive. In the universe, iron is only created is through the formation of supernovas and other massive stars.

At the beginning, the universe had elements like hydrogen and helium; but it had no iron whatsoever, so life could not have existed. The collision and explosion of hundreds of thousands of stars and galaxies— which happens because of gravity—produces iron. That is the same iron that runs through our veins. The fact that our veins carry an element created in faraway galaxies is difficult to comprehend, but there is no other explanation for that vital element within us.

Our planet appeared over four billion years ago. It was a very different place. At that stage, there was no oxygen on Earth. The atmosphere had as much nitrogen as it has now, but no oxygen. There was a lot of carbon dioxide. Only single cell organisms could live here. Some two odd billion years ago, there was an organism, called cyanobacteria, which could produce photosynthesis. That means they would take energy from the sun and would transform carbon monoxide into oxygen. They would synthesise light. Over a period of billions of years, those bacteria produced the oxygen we now have in the atmosphere and we can now breathe. The seas were produced first, and then the ozone layer. Without it there could be no multicellular life on Earth. And only after the Cambrian explosion could there be life on land. All those bacteria that lived so long ago were the origin of our lives. Cyanobacteria still exist in the plants we eat, which are sources of photosynthesis. We breathe in the same oxygen plants breathe out.

These few previous paragraphs attempt to provide a Western explanation for the beginning of life. It took a long time for sentience to develop within living beings. And a much, much longer time for cognition to become part of human consciousness.

～

HUMAN BEINGS ARE CURIOUS; scientists and philosophers are curious. That—many would say—is part of the cognitive component of our consciousness. Actually, purely sentient beings can be curious: cats are curious; many corvids, like crows and magpies, are curious; bears are curious. They are not inquisitive, though. What human beings do that differs from other animals is questioning. No other animal asks questions or gives explanations, or not that we are aware of.

Questioning is one of the expressions of curiosity. It is also an expression of doubt, of perceived possibility, or probability. Our questions have good and not so good qualities. The good qualities are the ones involved in science and philosophy: they are a part of our quest for knowledge. The bad qualities involve doubts about ourselves, fear or fears, anxiety, etc.

Questions in Zen, however, receive no answers (or no logical answers). Koans are famous for being unsolved puzzles. *"What is the sound of one hand clapping?" "Why did the Boddhidharma come from the West?"* That is, those questions are not well received. You can ask, but the answer will probably not be to your satisfaction. Zen Masters will tell you that you will not be able to find out about Zen through questioning, because sentience cannot be explained. It can only be demonstrated.

Zen doesn't explain. Anything. Zen accepts life as it is, without questioning or asking for explanations. Zen gives you an example of how things are: the flag flutters in the wind. Is it the flag that flies or the wind that blows? Neither. It's only your mind trying to explain something. Only your mind moves. The flag and the wind need no explanation. They just are.

∼

I CAN EXPLAIN A THEORY, an equation, a formula. I can explain a situation, a puzzle or a problem.

I cannot explain to you what being wet feels like, that lies in the realm of sentience. I can say that your skin is waterproof. I can say that if you jump into the water, you will go inside the water, but the water will not dissolve you; it is possible that you may swallow some, but it will not go into you otherwise. None of that will give you the experience of feeling wet. Diving or walking in the rain will. You cannot understand the act without experiencing it.

I can try to explain C minor to you. I can say that its key signature consists of three flats. Nothing will do, except hearing it. The same thing happens with the colour magenta. It is a light mauvish-crimson which is one of the primary subtractive colours. It was named after a battle that took place in Italy during the Napoleonic Wars. Explanations will mean nothing to you until you see the colour with your own eyes.

There are phenomena that cannot be explained with language or demonstrated through formulas. Ludwig Wittgenstein—arguably one of the greatest philosophers of the twentieth century—stated *"Whereof one cannot speak, thereof one must be silent."* (see chapter *Wittgenstein - understood by few*). Maybe he wasn't quite clear enough. Perhaps I could paraphrase or clarify the concept in the context of Zen: "Whereof one cannot speak, thereof one must allow sentience to take over.".

OUR SPECIES IS what biologists call an altricial species. Our young need an extremely lengthy period to be brought up; that means, they are reared biologically, socialised, and then they are cognitively educated to be able to live and thrive within human society, which is an artificial environment.

We are not born as cognitive beings. Cognition is artificially added to the animal layer of our consciousness. Cognition is transmitted culturally and intersubjectively. Our offspring are born exactly like those of all other animals. They acquire language and culture later, through their parents and the group. This has to happen every generation. Every human individual needs to be brought up and educated. The ability to think develops slowly from baby, to toddler, to child, to adolescent, and eventually, to adult.

But let us repeat here, and let us allow the concept to sink in: cognition is so artificial that we are not born with it. It has to be taught every generation.

What happens is that, after thousands of generations, it has become so ingrained in us that reality without cognition is something we find almost impossible to imagine. But we were like that as a species, and each one of us, individually, was like that.

I would not have been able to write this when I was a five-year-old. This is a fact, and this fact cannot be explained in any other way: cognition is a slowly-acquired—uniquely human—phenomenon. Sentience is innate, cognition is not. Sentience predates cognition individually and historically.

Saying that language is an innate human quality is a fallacy.

As a species, *H. sapiens* did not come equipped with a language template in the brain. That was the result of a feedback loop between language and biology. We have seen how language comprehension developed.

Our consciousness, then, has two components. I like to call sentience "evolutionary" and cognition, "meta-evolutionary".

Science and philosophy are cognitive. They are there to elucidate. They are equipped to deal with phenomena from a meta-evolutionary perspective. But neither of them can explain sensorial experience.

I see cognition as meta-evolutionary because everything that happened once humans became cognitive is different from what happened before then and from what happens to other species. Our societies do not work according to a Darwinian model or abide by Darwinian principles. There are institutions and rules that are exclusively human. As humans, we develop identities, a sense of what is good or evil, we create, we keep time, we produce art, we are adventurous, we have free will, ethics and morals. None of that appears in any other species.

A TIGER IS a tiger is a tiger. He is an iteration of tiger. He may feel as an individual, but he has no identity (other tigers do not call him Peter); he kills the prey he needs to eat and there is nothing wrong with that; he does not venture too far from his territory, and he does not produce or acquire anything that makes him different from other tigers. He lives in the present. He does not remember being a cub. He has no time, long-term episodic memory, or voluntary imagination, nor does he plan anything. He may apply tactics to hunt, but he never applies long-term strategy. Like other animals, he has some feelings or emotional memory. But he acts, primarily, out of habit and instinct.

The tiger is inwardly immortal because he has no idea of his own finitude. When he dies, he dies, but he does not ponder death. He does not stress, nor does he need to be competitive, except in a fight, where his life may be at stake. With a limited amount of cognition, he can solve basic problems.

We were a bit like him before we developed meta-evolutionary thought.

BUT LET us go back to cultural differences between East and West. Theologically and philosophically, East and West had similar beginnings. Buddhism was an offshoot of Hinduism, and Christianity an obscure cult that developed from Judaism. The results—however—were starkly opposing, irreconcilable, views of ourselves, of nature and the universe, both of which are partly correct. They are correct because of the hybrid nature of our consciousness.

The Hebrew Bible defined humans as something totally separate, above from the rest of creation. God put Adam in charge of all animals. Christianity, influenced by Greek thought, gave humans an immortal soul. When good Christians died, they went to Heaven, with God. They were individual demigods. That individuality was further emphasised by Martin Luther, who accepted no intermediaries between Man and God.

Humans could study a reality that was separate from them. Reality was objective. Nature was objective.

Secretly, astrology and alchemy gave birth to astronomy and chemistry, and other disciplines followed. Science crept slowly out of a nebula of superstition. Newton and Descartes—an alchemist and a magician—became the fathers of science and philosophy.

Things developed strangely. It was like a dance in which religion, art, philosophy, and eventually science, acted as the four bases of Western DNA, which intertwined to form the double helix of the culture. The West had opted for cognition.

AT THE BEGINNING, civilisation in the East—especially in the subcontinent—had grown out of all proportion: from small clans and trading settlements into major conglomerations of human beings. Cities like Mohenjo-Daro and Harapa had tens of thousands of inhabitants, which was hard to conceive at the time. The jump from hunter-gatherer groups to agricultural communities to cities had been fast and fraught with difficulties and problems.

There were castes and big differences in social stations; cities had the equivalent of gated communities, where the more affluent enjoyed their privileges. Many found that life in the city was stressful, or oppressive, or both. Competition and the need to acquire material things became too much and some either never integrated or eventually decided to drop out of the "rat race". There was a nostalgia for the old, more natural, life. Many reacted in a way very similar to the hippie movement of the sixties and seventies. Hippie interest in the East was no coincidence.

Fakirs, yogis and other mendicants appeared in the environs of big cities. Siddharta Gautama expressed the feelings of monks and other outcasts; he preached a reactive kind of non-attachment (to things or people). His "Middle Path" rejected ascetic life as much as carnal life and desires. Gautama had found a remedy against suffering. In large societies, cognition (and suffering with it) had been growing through the exponential use of thought, language and writing.

Gautama became the icon of the movement. He was declared the "Buddha" (the Awakened One).

BUDDHISM, born as a way of life, grew out of Nepal, meandered through India and then pushed into the continent where—from a philosophy—it became a religion with rites, scripture and all the other trappings of religion. By the sixth century CE, when the patriarch Boddhidharma travelled North, Buddhism reached China. There, it coexisted with Confucianism and Tao.

After nine years meditating before a wall, where his image became imprinted—or so the legend goes—Boddhidharma founded *Ch'an*, a meditation sect. It involved sitting cross-legged in what is known as the "lotus position". That position was called *"zazen"*, later shortened to Zen, in Japan.

～

How Zen entered Japan and became so influential in the country is a long story. Here, we can only say that, by the time it arrived, it had been distilled into the essence of Buddhism.

We know that, after countless millennia, the layers of human consciousness had become so enmeshed that it appeared impossible to peel one of them off. This is where Zen meditation holds its secret: meditation practice gradually dissolves cognition and exponentially heightens sentience; that is called *"satori"*. The fact that this is at all possible demonstrates that those strata, those layers, are discrete and have different natures.

We live stressing about the future and regretting things that we did or did not do in the past. When you focus your life in the present, you realise those problems do not exist.

When you practise Zen meditation and apply it to your daily life you will experience, at a certain point, something Masters call *"luminosity"*. That is the moment you realise that the activities of your body and your mind are not totally separate and you gain an incredible insight. Body and sentience are one and the same. During meditation, cognition and time slowly dissolve and, with them, your own identity disappears. You become wise; you are part of *"oneness"* again. Actually, you and *"oneness"* are the same. You and all other people (and nature) become one. Time does not exist. When that happens, no explanation is needed. Thought is gone.

HEZEKIAH

- WRITING THE SOUL OF A CIVILISATION

When you research the Book of Genesis, the *Tanakh* (or Hebrew Bible), the Old Testament or the New Testament, you discover many interesting things. You discover that we, as individuals, may not have souls, but nations do, cultures do. And they may not be immortal, it doesn't matter. The religious part is not that interesting, but Christian dogma and philosophy have influenced the views of the Western world all along in many transcendental ways.

In recent centuries, there were two important attempts at an exegesis of the Book of Genesis. One of them was by a French theologian called Isaac de La Peyrère who, in the 17th century, came up with a Pre Adamite hypothesis, i.e., that human beings had existed before Adam. That appeared to explain some of the inconsistencies of the Bible. He seems to have been forced to convert to Catholicism and eventually reneged on his views. Seventeenth century... of course.

The second attempt was by the famous 20th-century science-fiction author Isaac Asimov. Acknowledging that prior to modern historiography there was no rational version of history better than the Bible, he believed that the Book of Genesis had to be interpreted allegorically and that any literal interpretation of the Bible did not make any sense.

Both of their theories had some merit. I concluded that the Book of Genesis should not be interpreted literally, as that is a major insult to the intellectual capacity of those who wrote it. Like the rest of the *Tanakh*, Genesis is mythical in that it is a compilation of oral stories transmitted by the Hebrews around campfires and hearths during countless generations. It is a recording of what wisdom and knowledge the Hebrews had at the time it was written. Scripture.

It is easy to discover that the Hebrew Bible was written in very difficult times. By the eighth century CE, the old Kingdom of David and Solomon had been divided into Israel, to the North, and Judea, with Jerusalem as its capital, to the South. Hezekiah was the king at the time. The Assyrians had invaded Israel, which included ten of the twelve Hebrew tribes, and had dispersed its inhabitants. Many refugees from Samaria, Israel's capital, had flocked to Jerusalem, which created all kinds of problems in terms of work, famine, lawlessness, housing, etc. Judea needed a code of law, and Hebrew writing had been recently introduced and was being used on stelae and other official inscriptions. The King decided that it was the right time to produce a history, a religious record, a code of law—something that

Hammurabi had done centuries before—but now it would have the authority of the God of the Jews behind it. The Bible was born in times of crisis, but it was meticulously written, as one of its purposes was that it had to be used as the Law of the Nation.

INTERPRETED LITERALLY, The Book of Genesis does not make a lot of sense these days. But two things are important: it has to be interpreted correctly, and the reader needs to consider the time in which it was written and make allowances for that.

Adam was made from dust. Asimov says: *"The Biblical writers knew nothing of microscopic life, but dust is not a bad way of describing it, in the absence of knowledge"*. Yahweh was not meant to be a potter, but is there a better way of explaining the origin of human life to a nation of illiterate shepherds?

The literal version of what happened in the Garden of Eden —what most Christians believe—doesn't make a lot of sense, I said before. Let us see: they believe that God created Adam from dust. Then he created Eve from his rib. They were the first human beings. He put them in the Garden of Eden. He ordered them not to eat from a tree. He said if they did, they would surely die. Then the Devil, disguised as a snake, tempted Eve and convinced her that she should eat. Eve disobeyed God and persuaded Adam to do that as well. They both ate 'an apple'. They discovered they were naked and probably had sex, although that is not quite clear. God, in a fit of rage (?) expelled them from the Garden of Eden. He told Adam he would have to work and that he would die and turn to dust. He told Eve she would hate snakes and would give birth with pain. But then, after they ate the apple, they did not die, even when God said they would. He was wrong. God was wrong? Later they had two sons, Cain and Abel. Cain killed Abel. God became very angry (again?) and sent him away. He went away worrying that somebody would kill him. Who, if there was nobody else on Earth? Literally. He went East of Eden and built a city in the Land of Nod. Who were the inhabitants, if there was nobody else on Earth? He came back with a wife. Who, if there was nobody else on Earth? ... and so on and so forth. The literal Christian interpretation that Adam and Eve were the only human beings on the planet does not appear to make a lot of sense from a biblical perspective. Actually, it does not make any sense at all. Saint Paul said they had been created immortal and became mortal afterwards. Did he mean it literally? Nothing makes a lot of sense.

I would say it is quite probable that the scribes of King Hezekiah created a very beautiful, allegorical version of the myth. If we believe that Genesis should be understood allegorically, all of the inconsistencies of the Bible disappear. Of course, the presence of a deity in the picture is not quite what many people in the West—especially scientists and philosophers—would accept nowadays. But we have to take into account that the Bible was written almost three thousand years ago, and our understanding of reality has changed to a large extent.

Let us respect the ancient Hebrews as an intelligent nation. I would guess they were proto-Darwinian; they had some idea, some inkling, that human beings had evolved from other, more primitive, species. Being shepherds they had some knowledge that species could be improved for certain purposes by means of breeding. They knew about horses, donkeys, and mules. They probably knew about other primates.

WE HAVE SEEN that Aristotle and Plato, by using the word *zoon* when referring to human beings, appear to have implied that human beings were considered primates. They also used *"animal with thought/language"* (ζῶον λόγον ἔχον, *zoon logon echon*). It is quite possible that that was the natural understanding in the days of the *Torah* and before, when all the biblical stories originated: it was still understood that human beings had developed from apes. We were descended from animals.

Did the Hebrews believed this during Hezekiah's time, when at least part of the Bible was compiled? Or had the traditions and stories already lost their original meaning? If we put the question into chronological perspective, Aristotle lived in the 4th century BC and Hezekiah in the 8th century BC. Even allowing for the fact that they were located in two different geographic areas, the fact that there was a gap of four centuries between the two, and that Aristotle still considered himself a primate gives me a clear indication that people in Hezekiah's time were still aware of our animal origins, i.e., pretty much proto-Darwinian in their ideas.

Even though Judah was becoming slowly urbanised, they probably understood those traditions and legends with their original meaning. We have to remember that this was the early Iron Age and that living in caves, for instance, was not strange.

The majority of human beings were not city dwellers, even in the days of Jesus Christ. Many, like John the Baptist and his mother, or before then, Lot and his daughters, lived in caves in the desert. And of course, for many generations, Hebrews lived in tents like some Bedouin still do.

There are stories in the Bible that may allegorically refer to different stages of human evolution. Isaac, for instance, was tricked into giving his inheritance to Jacob, his youngest son. Isaac was blind and Jacob made him believe he was Esau, the eldest, by covering his hands with goatskin. Apparently, Esau, a hunter, was extremely hirsute. When Isaac touched Jacob's hands, he thought Jacob was Esau. The strange part of this old story is that Esau was so hirsute that his hands felt like the skins of goats. Allegory or exaggeration? The story refers to the period of the early settlement of Canaan. It has several readings, both literal and allegorical. In the days of Abraham and Isaac, towns were not the norm in Canaan, many people still lived in the open, in caves, or were nomads. The question is: When the Bible was written, did they think that all humans had always been totally developed? Or perhaps that there were some humans that were less developed than others? Maybe the story makes reference to the Edomites, Esau's descendants, who lived in current day Jordan, where Petra is. The *Torah* does not regard them as very advanced. The thing is: if it was reasonable to think that a human could be as hairy as a goat, the mental gap between human and animal was not as wide as it was to become later.

Stephanie Moser, an iconography specialist, attempts an explanation:

"Another biblical figure who conveyed a sense of the distant past was Esau, the brother of Jacob. Esau is described as being hairy all over and is frequently depicted as a wildman. The description of his physical appearance as hairy relates to his status as a hunter who lived in the wild. It also relates to his being the forefather of the Edomite nation, who may have been perceived by the Hebrews as hunters.". What Moser does not explain quite well is why hunters were supposed to be hirsute. Was there the impression that there were humans who were at different stages of evolution? The Hebrews, of course, did not understand concepts like genus and species. Maybe I am reading too much here, but it is possible.

As I state above, nowadays we know that in Europe, tens of thousands of years ago, Cro-Magnon individuals coexisted and interbred with Neanderthals. There were other closely-related, apes that coexisted with humans and were less advanced, like the Denisovans. Is it possible that in the Middle East the knowledge of those species had survived until biblical times? Again, it is possible.

Moser confirms that at a later stage, in classical times, barbarians were depicted as primitive, i.e., that outsiders or enemies were vilified that way:

"It was thus, at this early stage, that key icons for signifying the distant past were established, including the club, the animal skin, nakedness, hairiness and dark skin colour. These attributes effectively became visual symbols that played a critical role in communicating primitiveness and in separating non-Greeks and non-Romans. They signified an outsider or

barbarian status and summed up the qualities of the non-civilized existence."

So, even when this did not depict Romans or Greeks, there was an idea that there were humans who might not have been as evolved as others. Through St Paul, Christians may have inherited the concept, but understood it as a difference between Christians and pagans. Moser explains:

"In a general sense the visual icons developed in early Christian, medieval and Renaissance times functioned as part of a wider dialogue on how non-Christians were to be defined. This dialogue was inherently visual and relied on symbolic ways of conveying the primitiveness of a pagan existence."

Early Christians obviously never gave a second thought to the fact that, if there were human beings who were not very advanced or not very civilised in the Bible, it was possible that the ancestors of the whole of humanity were not very advanced or very civilised either. My conclusion is that it is very possible that ancient Hebrews had the intuition that, originally, humans had been apes.

LET us say the myth of Adam and Eve deals with the beginning of human consciousness. Then, it all makes sense. Imagine Adam and Eve as a couple of hominins who had probably walked away from the rest of their group and had started speaking and understanding each other. Of course, it's an imaginary scene. In real life, the beginning of human consciousness took tens of thousands of years.

Adam and Eve, being animals, would have known nothing about their future and their eventual death. In that sense, they would have been immortal, indeed. They would have lived day-by-day. But they were not the only hominins on Earth. They were the first ones who understood each other; who thought; who could decide whether their actions would be good or evil. No other animal species understood good or evil. In that sense—if we can imagine that such a couple ever existed—they were indeed the first humans.

After Cain killed Abel, he left his parents. He was scared other hominins could kill him. He found another group and returned with a wife from that group. Otherwise, there is no explanation for her existence. The wife had probably become totally human through language, and they taught their children to speak and think. Humanity commenced to grow and developed as a species.

APART FROM THE DEITY, if the Book of Genesis is interpreted allegorically, it makes a lot of historical, Darwinian, sense. The opposite is true of the literal, religious, interpretation of it.

Almost eight centuries later, Saul of Tarsus—aka St Paul—a Jew from the Greek diaspora, born in the Roman province of Cilicia (now Turkey), added some important details to the myth. As opposed to the rest of the followers of Jesus, who were illiterate, Paul could read and write and he was fluent in Greek and Latin, as well as Hebrew.

As mentioned before, Paul's interpretation of the myth was that before the Garden of Eden, Adam and Eve had been immortal. Of course, they were not yet human. They had no idea of their own finitude.

Paul—who was probably familiar with the Greek philosophers (especially Plato)— interpreted the myth with the new mentality of somebody coetaneous with Jesus, but with a much more sophisticated culture. According to Paul, what happened at the Garden of Eden was that Adam and Eve had received (a God-given) immortal soul (*psyche*). A suitable explanation for human consciousness at the time. The idea of the immortal soul is, of course, Platonic.

In the Hebrew Bible, the word most used to signify consciousness is *nephesh* (Hebrew נפש "breath of life"), i.e., the senses, or sentience. The word appears seven hundred and fifty times in the Tanakh.

The most important vocabulary addition of the New Testament is the word *psyche* (Greek: *ψυχή,* or "soul"), which is actually used one hundred and five times and was meant as the human mind with the inclusion of the second layer of human consciousness: cognition.

IN ANCIENT TIMES, people asked themselves whether the soul humans had was different from the rest of the other animals because they knew we had a special kind of consciousness. Well, we do. We have human consciousness.

As we have said, some Greek philosophers (Plato), in a most civilised fashion, believed humans had an immortal soul. St

Paul introduced an individual immortal soul based on Plato's but added details of his own to fit his Christian doctrine. Humans are better than and separate from all other animals. The Sadducees, who were the most important Jewish community, rejected the idea of the immortal soul then and Jews have rejected it ever since.

Aristotle thought we had souls, but mortal ones, like all other animals. Much later, St Thomas Aquinas went one step further and specified that only humans have immortal souls. He said that animals have souls that are mortal. These are the sort of decisions that plague Christian theology. However, in the West we had become separated from our fellow creatures in the animal kingdom at a much earlier stage. St Thomas Aquinas, in that respect followed what had been said before him. And every time you make a decision like that you have to explain the whys and the wherefores, and that is not always an easy task. What follows never, never, makes a lot of sense.

If we depart from the fact that all animals are alive—i.e., that they have a *nephesh*, according to the Hebrews—, then what Adam and Eve acquired in the Garden of Eden was what the Greeks called *psyche,* what St Paul imagined as the individual immortal soul, and what we now call cognition. If properly interpreted, then, the myth explains the acquisition of the two layers of human consciousness.

SCIENCE

SCHRÖDINGER

- INFORMATION AND LIFE HAVE DIFFERENT NATURES

"Reflection is the necessary copy or repetition of the originally presented world of perception, but it is a special kind of copy in an entirely different material.".

The World as Will and Idea —

Arthur Schopenhauer

Many books could be written about Schrödinger's contribution to knowledge. And—surely—many have. One of his finest achievements in physics was probably the wave equation—the behaviour of particles at the quantum level. A Nobel-prize-winning physicist, Schrödinger also ventured into biology, a science in which he collaborated in an important way with his discoveries regarding DNA, among other things. Until his death he continued to champion greater collaboration between physics, biology, and chemistry, especially to explain the emergence of life from inanimate matter. Some maintain that science has already solved some of those issues. In studying the origin of life, however, he proposed that living matter was ruled by aperiodic crystals, that is, it had a non-repetitive molecular structure. His were the first descriptions of DNA. He was, like Wittgenstein, a polymath.

SCHRÖDINGER BECAME close friends with Einstein and, like him, was trying to find a unified field theory. They exchanged copious correspondence on the subject.

In 1935 Schrödinger introduced the famous metaphor of the cat in the box.

The friendship with Einstein suffered when Schrödinger had the idea that a rotating mass would generate a magnetic field, which he published without consulting Einstein. Einstein told him that it didn't differ much from his theory. After that they stopped writing for three years.

Schrödinger thought a lot about his idea that a spinning mass produced a magnetic field.

As mentioned above, in 1926 he elaborated the mathematical formula of the wave function—the way in which the position of a particle can be described as a range of positions. On that basis, Heisenberg proposed the Uncertainty Principle regarding the position of particles, and then, Neils Bohr presented an idea that combined much of what physics had seen up to that point: the position of a particle can be described as a wave, and the wave is actually the probability of a position. By marrying that idea with the Uncertainty Principle, Bohr concluded that the properties of particles are totally random. Uncertainty is fundamental in the universe. We have already seen that Einstein opposed this idea, saying that *"God does not play dice with the universe."* Niels Bohr replied, *"Don't tell God what to do."* It may have nothing to do with God, but what has been proven so far is that Bohr was right about nature.

PARTICULARLY DRAWN to the most complex problems involving mind and matter, Schrödinger gave a series of lectures on consciousness at Trinity College in 1956. They were published under the title *"What is life?"*. This is what interests us the most about his work: he explored the nature of self-awareness and subjective experience; he delved into biological processes and their philosophical implications. And he provided some important answers.

LIFE APPEARED on Earth roughly 4.3 billion years ago, a huge number. The planet is teeming with it. Before trying to understand consciousness, human beings, or mammals, we have to understand what life is.

Schrödinger describes how life does not follow the Second Law of Thermodynamics. Life does not decay towards equilibrium. He says: *"Life seems to be orderly and lawful behaviour of matter, not based exclusively on its tendency to go over from order to disorder, but based partly on existing order that is kept up".*

Life is different from the rest of matter. Organic chemistry and biology follow their own rules. The vital components of living creatures are completely different from the rest of nature. Life is based on internal order. The essence of inanimate matter is the exact opposite: total chaos.

So, if life does not follow the same rules as the rest of nature, which is inanimate, and life—as far as we know—is exclusive to our planet, then life, apart from being fundamental (I repeat, something we cannot explain), could be also be a random phenomenon (which would confirm Neils Bohr hypothesis).

THE MOST IMPORTANT considerations on consciousness commence when Schrödinger discusses the main principles of Western philosophy and science, what he calls *"the principle of the understandability of nature, and the principle of objectivation."*. Nature can be understood, yes, but that has to be done by hypothesising that there is a real world around us, he says.

In order to understand that world, according to Western thought—he says— I have to exclude the subject from the picture. I become an onlooker. I do not belong in that world. Once I accept that, I encounter several problems, not the least of them, that my body (and the bodies of other people —other spheres of consciousness) are part of the real world.

So, the most important consideration here is that my own sentience is part of the material world.

Here, without saying it in so many words, Schrödinger prefigures Wittgenstein (the ineffability of some things) and our hypothesis of a "hybrid consciousness": *"a moderately satisfying picture of the world has only been reached at the high price of taking ourselves out of the picture ... finding our world picture 'colourless, cold, mute'. Colour and sound, hot and cold are our immediate sensations; small wonder that they are lacking in a world model from which we have removed our own mental person."*. So, we can only 'understand' the world but not our sensations, because they cannot be 'understood'. Cognition understands only cognitive phenomena.

Schrödinger goes on to say that he is repulsed by the idea *"that 'the world of science' has become so horribly objective as to leave no room for the mind and its immediate sensations."*.

Further, he states: *"Mind has erected the objective outside world of the natural philosopher out of its own stuff. Mind could not cope with this gigantic task otherwise than by the simplifying device of excluding itself — withdrawing from its conceptual creation. Hence the latter does not contain its creator."*. I take it that when Schrödinger says that mind erects the outside world of the philosopher out of its own stuff, he means that cognition only understands cognitive phenomena. Of course, he is quite aware that this is the construct of Western philosophy and science.

Schrödinger rejects the idea of the *'homunculus'*. We know our consciousness does not reside within our body, that little man that observes the world from between our eyes. The location of our mind is only symbolic, he says. There is, however, a ceaseless movement of neurones and electrochemical impulses, thousands and thousands of contacts every split second within our nervous system. So that, in order for us to understand, there are things that move within us. But then, as a quantum physicist, he remembers that the boundary between subject and object is really tenuous: *"We are given to understand that we never observe an object without its being modified or tinged by our own activity in observing it. ... The world is given to me only once, not one existing and one perceived. Subject and object are only one. ... for this barrier does not exist."*. Quantum physics and Eastern mysticism appear to agree on this.

The next lecture is 'The Arithmetical Paradox: The Oneness of Mind'. Here Schrödinger commences by saying: *"The reason why our sentient, percipient <u>and thinking</u>* ego is met nowhere within our scientific world picture can easily be indicated in seven words: because it is itself that world picture."* * (my underlining). The thinking layer of consciousness *is* the one that analyses the world. Sentience and perception do not analyse. They are tools of that analysis. They are used in order to provide evidence of fundamentals. They are the ones for which there is no explanation. The conscious mind —he is saying—cannot analyse consciousness. In any case, Schrödinger speaks here of the paradox of many egos and one world. The traditional Western perspective: individual minds are separate and self-contained.

Schrödinger then introduces the Upanishads and Eastern mysticism into the picture, and suggests the alternative: minds and consciousnesses are unified. Minds are only iterations of the one consciousness. He is right, of course, in terms of the oneness of mind, but he errs in not excluding cognition. Natural consciousnesses are one. Cognition is artificial. As a human creation derived from language, it does not share in their nature.

The other point he discusses in this lecture is the causality of consciousness: *"I find it utterly impossible to form an idea about either how, for example, my own conscious mind (that I feel to be 'one') should have originated by the integration of the consciousnesses of the cells (or some of them) that form my body, or how it should at every moment of my life be, as it were, their resultant."*. His intuition seems logically correct to me. Consciousness appears to be a holistic phenomenon. Also, from an evolutionary perspective, organs follow behaviour, not the other way round. *"The single nerve-cell is never a miniature brain"*, he says. I would venture that isolated lab-grown brain organoids (which didn't exist in Schrödinger's time) will never acquire full consciousness. Sub-minds are monstrous, as are plural minds. The first ones are artificial, the latter ones have never existed. There are separations of the sensorium into different areas, that's all. That is Sherrington's paradox. The nervous system operates on the basis of the integration of many sub-systems.

Schrödinger submits two paradoxes of consciousness, one internal and one external. He then proceeds to indicate how to reach a solution: *"I submit that both paradoxes will be solved (I do not pretend to solve them here and now) by assimilating into our Western build of science the Eastern doctrine of identity."*. That is exactly what I am attempting to do here. The new element I introduce is the duality sentience/cognition.

Both, Western and Eastern schools of thought are partially correct because human consciousness has two layers. The cognitive layer asks questions and receives some answers (through religion, philosophy and science). As we mentioned before, some may be true, some not so true (I do not want to judge). As for the sentient layer, it asks no questions and receives no answers. It has no need because it is one with nature.

He continues: *"The [cognitive] model is colourless and soundless and impalpable. In the same way and for the same reason the world of science lacks, or is deprived of, everything that has a meaning only in relation to the consciously contemplating, perceiving and feeling subject. I mean in the first place the ethic and aesthetic values, any values of any kind, everything related to the meaning and scope of the whole display. All this is not only absent but it cannot, from the purely scientific point of view, be inserted organically. ... For anything that is made to enter this world model willy-nilly takes the form of scientific assertion of facts; and as such it becomes wrong."*.

Briefly touching upon ethical considerations, Schrödinger tells us that life is valuable, but that nature does not value it. Nature does not issue ethical judgments. *"There is nothing good or bad but thinking makes it so. No natural happening is in itself either good or bad, nor is it in itself either beautiful or ugly. The values are missing, and quite particularly meaning and end are missing. Nature does not act by purposes."*. We are the only ones who ask 'what is good and what is bad'. He is telling us that we are the witnesses and, he adds: *"The show that is going on obviously acquires a meaning only with regard to the mind that contemplates it."*.

The last of Schrödinger's lectures, *The Mystery of Sensual Qualities* is perhaps the most revealing. He appears to concur with the notion of a clear separation of the components of human consciousness: *"... all our knowledge about the world ... rests entirely on immediate sense perception, while on the other hand this knowledge fails to reveal the relations of the sense perceptions to the outside world, so that in the picture or model we form of the outside world, guided by our scientific discoveries, all sensual qualities are absent."*.

Schrödinger explains in all kinds of scientific detail how the senses operate: the physicist's idea of yellow light is that it consists of *"transversal electro-magnetic waves of wave-length in the neighbourhood of 590 millimicrons."*. Notwithstanding that *"The sensation of colour cannot be accounted for by the physicist's objective picture of light-waves."*. The bottom line is that the explanation of the physicist does cannot provide the sensation. Could a physiologist do better? Schrödinger doesn't think so. And the same thing that happens with yellow happens with sweet taste, musical notes, touch, hot and cold, smell, taste, etc. Nothing works. The paradox is that *"... the direct sensual perception of the phenomenon tells us nothing ..., yet the theoretical picture we obtain eventually rests entirely on a complicated array of various informations, all obtained by direct sensual perception."*.

LET'S say energy and matter are "fundamental" notions. Fundamental notions are notions we cannot explain, basically because we cannot fully explain how the universe started. Energy and matter came with the original package.

How? We don't know. Above, I propose that life is also a fundamental notion, albeit the result of the random behaviour of particles.

Schrödinger postulated that consciousness was also a fundamental phenomenon. I believe by that he meant just sentience. The principle is an important one in Eastern schools of thought: the interconnectedness of our (apparently plural) consciousnesses and the underlying oneness of nature. Schrödinger, however, believed that a fundamental aspect of the universe was a web of interconnected energy 'and information'. Eastern thought proposes the oneness of sentience but excludes cognition (i.e., information) from it. Again, I submit that cognition and information are exclusively human phenomena, and artificial to boot. Cognition and information cannot be included as part of any fundamental phenomenon.

I WOULD GUESS that sentience (which is an exclusive quality of life) is as fundamental as life. As such, it is part of the oneness of nature. It is also within us, who are living beings. We are part of it. Our senses do not need communication or explanation because they are part of the oneness. They (and we) *are* the oneness. Sentience does not need time nor does it need quantities. It has no numbers. It is one.

Cognition, on the other hand, is a by-product of meaning, which became language (an 'invention' according to Everett), was retrofitted into our consciousness. It is only subjective. It needs to understand, it needs to explain, it needs to communicate. Cognition comes with ethics and morals and beauty and judgement. Cognition comes with many other qualities that make us unique, and it is also the inquisitive component of human consciousness (I was going to write 'human nature', but—if I am correct—cognition is not natural). It has allowed our species to grow exponentially, however, it has allowed us to witness the universe, and to understand many natural phenomena.

The origins of the cognitive phenomenon, as explained above, came with language. After countless generations of combining sounds and adding meaning, a hominin uttered a combination of sounds with complex meaning (syntax). The interlocutor understood the meaning. Bingo. Humanity.

NEUROSCIENCE

- FUTILE ATTEMPTS AT MEASURING EXPERIENCE

"Finally, any attempt to deny that your immediate awareness has outside sources is incoherent, because this denial cannot be thought or stated except through the conceptual and linguistic resources provided to you by your social heritage and your cultural and biological evolution. Even if you were the last surviving person of a universal plague, your mind and your capacity for meta-awareness would be inherently social."

- Adam Frank, Marcelo Gleiser & Evan Thompson

Currently, there are several attempts at defining consciousness. Also, the perspective of the Cartesian dualistic framework (that recognises human nature as a combined physical and metaphysical system) is now generally considered a scientific taboo.

Even though mind and matter are clearly two distinct ontological entities, only monist notions are currently acceptable to most in the scientific community. The majority of neuroscientists only study the brain as a physical, individual, entity. The paradigm purposely ignores the existence of culture as a valid component of consciousness. Sadly, it is a solipsistic view of the world, where only the material, individual aspects of the brain are considered "mind".

Again, from my humble perspective, any study of consciousness must include, at least, cognition (with language and culture subsets), or else you are actually excluding full human consciousness.

From an etymological point of view, objecting to the use of the term "consciousness" to mean "sentience" alone is easy. The term has Latin origins: "Scientia" means "knowledge" and "cum" means "with". "Conscientia", thus, implies "shared knowledge". The knowledge we share, socially or culturally. Sentience is exactly the opposite. Understanding and knowledge are part of cognition, not of sentience.

Scientific reductionism attempts to circumscribe consciousness within a locus or loci in the individual brain. That is impossible: the term implies something intangible that allows communication with other human individuals. Using "consciousness" interchangeably with "sentience" needlessly isolates it and adds unnecessary complexity to the discussion.

At this stage—in the words of a neuroscientist—neuroscience prefers to study *"a 'VW beetle' in order to eventually be able to explain a 'Tesla'"*. The analogy is not an apt one. In reality, neuroscientists would like to explain a rocket as those used by NASA or Starlink through the study of a Ford Model T. The evolution of the rocket involves the quantum leap taken by the Wright brothers, the invention of the airplane. You cannot explain full human consciousness in evolutionary terms.

A REDUCTIONIST, Dr Kevin Morris, from Tulane University, author of *"Physicalism deconstructed"*, holds the view that the brain *is* consciousness. Apparently, there would be nothing over and above the physical world. That view is very difficult to understand for anyone who holds the commonsense view that ideas, and knowledge (and communication to a certain extent) are non-physical entities.

At an interview, however, Morris stated he believes in a more expansive concept of physicalism (?) than other philosophers: *"Physics, austere mathematical physics, describes things relationally, in terms of dispositions, in terms of how well one thing interacts with another, and you might think—for reasons that that have nothing to do with consciousness— that there must be more to the world than just relations or dispositions."* That view appears to contradict the notion that the brain *is* consciousness. Things do not get much clearer from physicalists, other than that anything metaphysical does not exist.

When Newton came up with the law of universal gravitation, he explained that particles attract each other with a force proportional to the product of their masses. He did the same with the motion of objects: an object may be at rest or in motion. If it is in motion, it will keep its speed constant and in a straight line unless an external force influences it. He also explained acceleration and the different forces that act on an object. This happened in the seventeenth century. He did not explain why these phenomena happened. In the twentieth century, Einstein refined the notion of gravity, not as a force, but as the curvature of space-time caused by an uneven distribution of mass. It is now the twenty-first century and we still don't know why these things happen. They are called "fundamentals". There are many fundamentals. These are phenomena whose origins are beyond knowledge.

As mentioned above, we have an idea of when life started on Earth. We still cannot explain why it behaves the way it behaves. Normal laws of physics do not apply to biology. Why is that? Maybe life is another fundamental. Sentience is a phenomenon that comes with life. There is no evidence of any inert object, like a stone, being sentient (or conscious for that matter). I would say—contrary to some theories of consciousness—that stating that inert matter might be conscious, is rather nonsensical.

Neuroscience keeps on studying individual brains in the hope that it will find the place where consciousness emerges. There is no such place. Our brains have a neocortex, which is the product of our cultural consciousness through a feedback loop. Our neocortex appears to have grown from the moment we acquired our distinct human consciousness, i.e., its origin, and spectacular growth, seems to be cultural. In any case, even if neuroscientists find a physical place, how will the study explain the 'emergence of consciousness'?

Individual human consciousness is not purely physical. It does not emerge from the brain or in the brain; it's the other way around: neurons and genes are created as a result of behaviour.

Apart from the neocortex, the human hippocampus has grown, no doubt, because we have more cognitive functions related to memory, whereas the memory areas to do with emotion have decreased. That is noticeable when compared with those of other primates.

A 2020 article by Rogers Flattery *et al* gives an account of the proportional growth of the learning and long-term-memory-related hippocampus area in humans as opposed to emotional memory:

"The hippocampal formation is important for higher brain functions such as spatial navigation and the consolidation of memory, and it contributes to abilities thought to be uniquely human... In the context of prior investigations of rhesus monkeys and humans, our findings indicate that, in the hippocampal formation as a whole, the proportions of neurones in CA1 and the subiculum progressively increase, and the proportions of dentate granule cells decreases, from rhesus monkeys to chimpanzees to humans."

What happens is that the neocortex and the hippocampus are part of an information system that individuals share with the rest of humanity. Collective consciousness was the effect of culture and—amazingly, and artificially—we grow from infancy to become integrated into that common culture.

In another recent article, Ben Turner describes the findings of a study conducted by physicists at the University of Sydney. The study discovered certain mysterious wave patterns in the neocortex. Turner reports:

"The wrinkled, outermost layer of the brain —known as the cerebral cortex— manages many of the mind's most complex tasks, such as memory, attention, language, perception and even consciousness itself... Yet neuroscience has mainly ignored the cortex itself and instead traditionally focused on the connections and interactions between neurones (the brain's nerve cells) to determine how the wrinkly organ functions."

What he is saying is that neuroscientists appear to ignore the outer cortex—where most of human consciousness resides—and concentrate on solely biological functions. Of course, this physicalist bias—that concentrates not just in the individual brain, but only part of it—will not help neuroscience in its quest for human consciousness.

RIGHT NOW, there are many theories of consciousness but there is no universally agreed definition of consciousness. The trend is definitely monist, physicalist, and reductionist.

From a lay person's perspective, the need for a clear and agreed-upon definition of consciousness is evident. To study human consciousness effectively, one must delineate its boundaries, which should encompass cognitive, cultural and linguistic aspects in order to fully include the 'human' aspect of that consciousness. A purely reductionist approach falls short of explaining the intricate exchanges and transformations of information that occur during cultural communication.

DARWIN & WALLACE

- DIVERGENCE OVER HUMAN EVOLUTION

"A great stride in the development of the intellect will have followed, as soon as the half-art and half-instinct of language came into use; for the continued use of language will have reacted on the brain and produced an inherited effect; and this again will have reacted on the improvement of language. As Mr. Chauncey Wright has well remarked, the largeness of the brain in man relatively to his body, compared with the lower animals, may be attributed in chief part to the early use of some simple form of language,- that wonderful engine which affixes signs to all sorts of objects and qualities, and excites trains of thought which would never arise from the mere impression of the senses, or if they did arise could not be followed out. The higher intellectual powers of man, such as those of ratiocination, abstraction, self-consciousness, &c., probably follow from the continued Descent of Man -improvement and exercise of the other mental

faculties."

Descent of Man -

Charles Darwin

In *Descent of Man*, it is possible to appreciate that Darwin himself could see the effect language (and culture, of course) had had on the human brain. He could see the feedback loop that had acted on the adaptive brain and produced the growth of the neocortex. Darwin guessed that—at a certain point in time—there had been a massive leap (meta-evolutionary—I would add) that placed *H. sapiens* well beyond sentience and any other species.

It is not that cognition is entirely dependent upon language, but it could be said that there is a pre-linguistic basic degree of cognition and a linguistic one.

Our minds appear to operate at two different levels to achieve two different ends.

One layer of the mind—sentience—is there to keep our individual bodies alive. Like all other mammalians, we need our senses to see, smell, hear, taste and touch. They allow us to move unimpeded in our surroundings, enjoy food or music; see, smell or hear predators, enemies, or sexual partners; and recognise familiar shapes and textures with our fingers or toes when we cannot see them, among other vital activities we require to survive as individuals. That layer is totally related to biology, totally physical. It is a result of evolution and all mammalians like us have it.

The biological part of our consciousness, which is also the 'emotional' one, concentrates mostly on survival, it regulates body temperature, breathing and heartbeat, for instance. That is the animal (as opposed to human) component. Whenever there is trauma, or danger, the mind initiates its response: "fight or flight". All kinds of chemicals flood our body, from adrenaline to cortisol. Rational thought becomes absent. The subconscious is also at play, the same thing that happens in dreams.

A strange phenomenon has often been reported during these "fight or flight" events in human beings: time appears to become dilated. What actually happens is that in those situations sentience takes over, with the consequence that time virtually stops, as it does not exist for the senses.

Our main concern here is the addition of the second component—with it, our consciousness really becomes human consciousness. Thought occurs within the individual and is —in the words of George Steiner: completely *"impalpable"*. The physicalist, Neo-darwinian study of the human mind occupies itself with sentience.

At the beginning of this chapter, we saw that Charles Darwin had guessed that language greatly influenced cognition. He could surmise that dualism provided a good explanation for consciousness. Unfortunately, the nineteenth century was a period when the only option he had was materialism or religion. With no other rational possibility, he chose materialism. He often repeated that human beings were too proud to believe anything but Creation: *"Man in his arrogance thinks himself a great work, worthy the interposition of a deity, more humble & I believe true to consider him created from animals"*.

His biographers—Adrian Desmond and James Moore—tell us he adopted an unbelievable explanation for consciousness, only understandable in a person of Darwin's intellect because of the times in which he lived:

"[Professor John Elliotson's] stock provocation was that the brain exudes thought as the liver does bile. It was Darwin's 'bon mot' exactly. 'Thought, however unintelligible it may be, seems as much function of organ, as bile of liver'. But Darwin's goading had a sting that even Elliotson's lacked. Everyone accepted that gravity was an intrinsic 'property of matter,' no one made a spiritual adjunct. So 'Why is thought' not seen as 'a secretion of [the] brain' in the same way? 'It is [because of] our arrogance, it is our admiration of ourselves'."

.

What neuroscience proposes nowadays—not in the nineteenth century—is a similar notion: consciousness emerges from neurones.

The mainstream Neo-Darwinian scientific world attempts an explanation of consciousness solely following a gradual accumulation of random genetic mutations through natural selection, heredity and isolation. We believe that is wrong on two accounts:

1. evolution can also occur through alternative mechanisms like spontaneous mutations, i.e., genetic insertions, or symbiogenesis;
2. consciousness in humans has a cultural, artificial, component that cannot be explained through physical evolution.

Why can't biological evolution alone explain human consciousness? Are there any other indications as to why not?

Well, we all know that mutations mean change. Because of their randomness not all mutations are beneficial to an individual or to a species. Very often the result of a mutation is exactly the opposite. In those cases, individuals who have mutated die. What helps beneficial mutations is natural selection. Natural selection produces mutant individuals that are better suited to survive in a given environment. This is probably Darwin's crowning discovery. Random mutation needs to be combined with natural selection for the species to thrive.

Humans are the result of countless biological mutations that helped them survive up to a certain point. That point was the beginning of humanity, the beginning of language and culture. Then—because of language and communication—the mutations stopped being individual and became largely cultural. Certain cultures progressed further through beneficial mutations. Groups change and improve.

Biological mutations—which apparently (?) continue to occur in our species—are not helping individuals to survive any more. Human institutions provide against the survival of the biological fittest. Human culture opposes the "law of the jungle".

Human individuals belong in groups, and those groups provide support to individuals. There are rules that each culture imposes because groups without rules perish. The result is that rational individuals communicate within their own culture, and physical fitness—although desirable—is not necessary for the individual to survive and thrive in society. What is necessary is intelligence, together with other qualities.

The emergence of human consciousness does not fit easily within Darwinian evolution. That is because it has no explanation within that scheme. It was a protracted process that climaxed with language and thought. Part of it could be considered from within the evolutionary process. Much of it, especially its culmination, with logic, reasoning, modesty, human abstract institutions, has to be considered meta-evolutionary. Wallace knew it. Darwin suspected it. There is no physical explanation. The only explanation is collective (cultural and dualistic). The non-physical nature of human consciousness cannot be ignored.

THE MAIN REASON for the scientific bias towards physicalism and reductivism is that biology is a physical science and no explanation can be found for collective consciousness within its bounds; but there are other reasons that contribute to the rejection of dualism.

Neuroscience has currently adopted a monistic approach. That has a logic to it. While Descartes, a philosopher, believed that human beings have a dual nature—what in those days was considered a body and a "spirit"—, neuroscience has devoted itself to studying only the biological side of that nature and denies the existence of a "spirit", or a "soul". Well, let's make it perfectly clear: nowadays we know that what drives human beings is not a soul, but our mind (*psyche*), our human consciousness (both, sentience and cognition). It obviously exists and it is intangible; it does not emerge from neurones. Neurones emerge from it.

The undeniable, unfalsifiable, "news" is this: humans are not born with our full human capacities completely developed. We are born only equipped to be fully human. But, at the time of birth, we can neither speak nor communicate. We do not understand our parents, our siblings, our relatives. They love us and care for us. And we can do nothing but allow them to love us and care for us. With speech, with language, with communication, comes cognition; all of them are skills that humans need in order to live in society, within a culture; and we are social beings. We need to live in society and we need society to survive. A bear lives on its own. We cannot. We possess a highly adaptive capacity to learn from our experiences, but this capacity would not have survived prehistory without the cumulative generation-to-generation transmission of knowledge. Among humans, information is transmitted from one generation to the other, but also concurrently, within the same generation.

RECENT EVENTS DEMONSTRATE the chaotic situation in which the neuroscientific world finds itself. As we have already explained here, there are many theories of consciousness but there is no universally agreed definition of consciousness. The trend is definitely monist, physicalist, and reductionist.

Some philosophers of science—like Alva Noë—however, reject monism and describe the *"human being as a kind of bio-cultural phenomenon"*. Totally agreed.

Another central problem is that of feeling, or "qualia" if you like: the subjective phenomena that constitute experience. That is, how, as individuals, we feel pain, enjoy music, or go through other sensations. The question posed by Chalmers that originated the "hard problem of consciousness", relates to how you can explain qualia in terms of neural correlates in the brain.

FROM MY PERSPECTIVE, things will become clearer: at a certain point, one of our hominid ancestors was intelligent enough to understand a message, a request—or probably an order—from another primate. He or she was able to connect sounds, emitted by the latter, with a relatively complex meaning. He or she understood. That was all; that was the beginning of cognition within human consciousness. It was the beginning of humanity, which was cultural. And linguistic. The Adam and Eve phenomenon. The repetition of that act, probably through many generations, created neurones that ended up being gyri in the cerebral cortex.

It was a giant leap. After that moment, humans became the only animals with a complex recursive language that ended up involving present, but also past and future: i.e., possibility, long-term memory, and voluntary imagination, which other animals appear to lack or have in smaller degrees. Information and knowledge grew in sophistication. At the same time, the cerebral cortex expanded exponentially. Human brains needed larger craniums to contain them and women gave birth with pain. But above all, humans were able to co-operate to a much larger degree and excelled at functioning as social beings in greater numbers.

Through religion, which is nothing but another quest for knowledge—an old form of science—the ancient Hebrews explained how before language there was no humanity. As I hinted above, (non-religiously, I emphasise), the myth of Adam and Eve in the Garden of Eden poetically addresses issues like communication, guilt, punishment, work and, basically, the beginning of consciousness, self-awareness and identity. Two hominins, one male and one female, become human. That was the myth. We are now aware that, from that moment on—not as God warns in the myth, but in reality—human lives would change dramatically. We learned to live in societies increasingly larger. And we developed something that was non-physical: a collective consciousness. A corpus of information shared by the collective, by the culture.

What should be obvious to science by now—and does not appear to be—is that human consciousness cannot be found within just an individual brain. Cultural and linguistic changes and exchanges are common, and normally consensual. Human consciousness is partly linguistic and cultural.

What is not evident either is that language and culture are a "black swan" phenomenon, as I said; the fact that an ape understood another ape was probably not that strange or not that far beyond normal biological evolution: other animals communicate. But if we deem that act as the beginning of language, culture and society, it was definitely meta-evolutionary.

With the advent of humanity and cognition came other phenomena that were definitely not somatic, such as memory, voluntary imagination (not dreaming), communication, co-operation, creativity, and adventurousness, i.e., collective consciousness. The latter is a non-physical entity that lives within a culture and interacts with the individual. A culture is more than the sum of the individuals that are part of that culture, it involves a history and a possible future, it is dynamic: it is synergetic.

Alva Noë (*The Entanglement*) succinctly describes what the science of consciousness needs to do now:

"Modern biology achieved its full explanatory power, its ability to account for life, its variety, and origins, thanks to the Original Synthesis, that is, the integration of Darwinian evolution with Mendelian genetics, but also with the new molecular biology that came of age in the mid-twentieth century. But if we are to explain the 'human mind', it is now believed by many, we need a New Synthesis, that is, we need to join biology, so understood, to the theory of cultural evolution". Yes.

I repeat—the way I see it—human consciousness has two layers: a basic, biological one, individually represented by the reptilian and the mammalian parts of the triune, adaptive, brain; and a cultural one, represented by the cortex. The latter appears to have grown as a direct result of social and linguistic interactions. The layers are integrated, but they are definitely discrete, as they originate in different phenomena: one of them is evolutionary, and the other one, meta-evolutionary, as explained. That also means that they have different natures. What Noë calls 'cultural evolution' I call 'meta-evolution'.

Yuval Noah Harari's account of scientific progress in the field of consciousness is devastatingly clear:

"To be frank, science knows surprisingly little about mind and consciousness. Current orthodoxy holds that consciousness is created by electrochemical reactions in the brain and that mental experiences fulfil some essential data-processing function.

However, nobody has any idea how a congeries of biochemical reactions and electrical currents in the brain creates the subjective experience of pain, anger or love. But as of 2016, we have no such explanation, and we had better be clear about that."

More than eight years have passed and neuroscientists continue on the same path.

The title of Harari's second book (*Homo Deus*) suggests we have become deities.

The way I see it, our species, *H. sapiens*, could be aptly named *Homo Creator*, as we have collectively forged an intangible entity—human consciousness—within our cultures and societies, which unites us as witnesses of the universe. Our religious ancestors believed in the Holy Spirit. Today, we have manifested a universal hologram of it in the digital realm.

∼

BUT LET us go back to Darwin. Why did it take so long for someone to develop a theory of evolution? The idea, actually, had been around for many centuries. We believe the Hebrews were proto-Darwinian. A close analysis of the Book of Genesis confirms this. They appear to have understood that humans had evolved. Certainly, there were many others who guessed the possibility. Hobbes was certain that our human ancestors could not foresee their own death. The Comte de Buffon, a French naturalist, saw the relationship between humans and other apes.

In his poem, *"De rerum natura" ("The way things are")*, Lucretius, the first-century CE Roman poet, imagines how the first humans lived:

> "Of sun withdrawn forever. But their care
> Was rather that the clans of savage beasts
> Would often make their sleep-time horrible
> For those poor wretches; and, from home y-
> driven,
> They'd flee their rocky shelters at approach
> Of boar, the spumy-lipped, or lion strong,
> And in the midnight yield with terror up
> To those fierce guests their beds of out-
> spread leaves."

What is certain is that Erasmus Darwin, a physician, and Darwin's grandfather, had written about the possibility of evolution long before his grandson. What Charles Darwin contributed to the theory of evolution—apart from an incredible amount of research that took most of his life—was something very innovative: the idea of natural selection, i.e., the *"survival of the fittest"*. Of course, that does not apply to human civilisation, because, as we know, civilisation is a result of cognition, which was meta-evolutionary.

The reaction of the general public at the time of *Origin*'s publication was one of derision. How could anybody say that we are descended from apes? The brunt of the criticism came from religious conservatives and some scientific sceptics. Darwin was disheartened but his main concern was the reaction of the scientific community. Among the latter, one of the worse was that of Adam Sedgwick, Darwin's geology teacher: *"I have read your book with more pain than pleasure. Parts of it I admired greatly; parts I laughed at till my sides were almost sore; other parts I read with absolute sorrow.".* John S Henslow was a Cambridge academic, friend and mentor of Darwin's; his first criticism was milder but it still hurt Darwin deeply *"... [the book] no doubt contains much legitimate inference—but it pushes hypothesis (for it is not real theory) too far.".* Eventually, Henslow accepted and even promoted Darwin's work.

~

IN MANY WAYS A RENAISSANCE MAN, Alfred R Wallace was a jack of all trades within the scientific world; he had been an amateur entomologist—like Darwin—, but also a biologist, and an anthropologist among other things. He was, much more than Darwin, a scientific polymath. He had read *The Voyage of the Beagle*, and some of Lamarck, Saint-Hilaire and Erasmus Darwin's ideas on evolution, which apparently inspired his own ideas on the subject. He travelled extensively in North and South America and Australasia, especially in Indonesia, Malaysia and Singapore. But, most importantly, Wallace came up with a fairly comprehensive theory of evolution and natural selection almost simultaneously with Darwin. He corresponded with Darwin and—

it seems—his letters prompted Darwin to publish *Origin* earlier than planned in order to claim precedence. In the mid 1850s, Wallace wrote a paper (*On the Law which has Regulated the Introduction of New Species*) which did not specifically deal with evolution, but was evidence of the way his ideas were progressing. In 1858 he sent Darwin his seminal paper *On the Tendency of Varieties to Depart Indefinitely from the Original Type*, which coincided almost exactly with Darwin's ideas.

But there were differences: while Darwin emphasised competition within species, Wallace focused more on the environment and how species had to adapt to their local area and diverge from the rest. This may not appear as an important distinction, but it partly led Wallace to differ from Darwin on human evolution. Wallace saw human evolution in terms of stages: first, bipedalism, and then, *"...the recognition of the human brain as a totally new factor in the history of life"*. Among the evolutionists of his time, Wallace was the first to notice that the evolution of the human brain resulted in the evolution of the rest of the body being redundant. His ideas on how human societies and cultures evolved were fairly advanced, as opposed to Darwin's, who had not even considered much of that area.

Unfortunately for Wallace, the fact that he was a spiritualist, played against the scientific approval of his theories. But I believe he was right in considering the development of human consciousness as a phenomenon separate from natural selection. Human consciousness—he believed—could not have had solely physical causes, i.e., it had not been produced by evolution.

LANGUAGE & CULTURE

WHORF

– HOW LANGUAGE INFLUENCES THOUGHT

"It seems plain and self-evident, yet it needs to be said: the isolated knowledge obtained by a group of specialists in a narrow field has in itself no value whatsoever, but only in its synthesis with all the rest of knowledge and only inasmuch as it really contributes in this synthesis toward answering the demand, 'Who are we?'"

Science and Humanism ——

Erwin Schrödinger

---

*H*uman individuals communicate via a mutually intelligible language. That means that, to communicate complex notions, individuals need to use the same language and, to a large extent, understand the culture of that linguistic community.

In order to speak any given language with some degree of fluency, one has to understand the culture. That is an important and unavoidable fact.

SOMETIMES WE HEAR that some concept has been "lost in translation". That is because the translator or interpreter did not comprehend the meaning hidden in a few words that appeared out of context.

Idioms and sayings are good examples of how a knowledge of the culture is essential to comprehend meaning. They are instances in which usage has maintained certain terms that may have become cryptic to the foreign ear. The other day I was telling my wife in Spanish that we had to do something "Just in case...", I said "No sea cosa que...". The literal English translation of that phrase is: "Not be thing that...", which makes no sense to an English speaker, or to any non-Spanish speaker for that matter. I could have made it worse. I could have used a phrase that is more common in Spain than in other Spanish-speaking countries: "Por si las moscas...", literally, that means "For if the flies...". The same thing happens when you translate literally an English idiom like "To be under the weather" into any other language. It does not make sense.

A judge once told me to repeat to him what the defendant was saying "word by word". I explained that if I did that, he would not understand. He insisted: "literally", he said. I repeated word by word something the judge could not understand. Then I explained it in English. It sounded totally different. The judge understood. He understood both messages, what the defendant was saying, and the fact that languages are not interpreted or translated literally.

∽

REMEMBER THE MOVIE "DANCES WITH WOLVES"? That was the name the Sioux had given the protagonist of the movie. An English native speaker would never have thought of a name like that. It's a sentence (i.e., it includes a verb). It sounds strange. English has a predominance of nouns (it is a more static language), whereas languages indigenous to the Americas, like Arapaho, place the emphasis on verbs (they are more dynamic than their European counterparts). The word for 'cement' in Arapaho is roughly 'it has hardened', and the term for 'chair' is 'the place where you sit'. In these two cases the language emphasises function rather than feature. It is something that happens instead of a static characteristic of the subject. These differences permeate languages and inform the way a speaker sees the world.

∽

ALL OF THE above examples of cultural influence on language may appear anecdotic, but they point to one underlying fact. There is a feedback loop between language and culture. And those two influence the thought processes of the speaker of any given language. The following chapter, basically about how Averroes misinterpreted Aristotle, provides a clear example of the chasm between some cultures. There are terms that are incomprehensible to speakers of other languages, and that is because the institutions that exist in one culture may not exist in the other.

IN THE 1930S, Benjamin Lee Whorf, a student of linguistics at Yale University, under Edward Sapir, was working on a grammar of the Hopi language, when he made some interesting discoveries. Hopi verbs do not have tenses, only aspects. What happens is that, in Hopi, time and space receive treatments that are different from those of other languages, and those treatments are not readily understandable to non-Hopi speakers. What Whorf discovered was that Hopi speakers did not appear to have a concept of time, or—if they did—theirs was a *sui generis* one. That implied that their thought processes differed from those of the speakers of other languages and—given the close relationship between language and thought— that probably the Hopi language influenced their thought processes.

Whorf concluded that language influences thought. The conclusion that may be drawn is that language is not universal. Another way of looking at it is that language and culture are not natural but artificial. They are human-made.

THERE ARE many examples of words that translators and interpreters need to explain. We saw that the Hopi language does not have a word for "time", that its verbs have no tenses.

Pirahã, among many other languages, does not have numbers. How can you explain the function of a wristwatch in Hopi, or that we need to divide equally the fish we have caught to a Pirahã speaker? We have forty-nine fish. We need twenty-four and a half each. It is almost impossible without showing them what to do.

As I mentioned above, in view of those rare characteristics of the Hopi, Whorf came up with a hypothesis later known as the Sapir-Whorf Hypothesis, or the Whorfian Hypothesis, or Linguistic Relativity: a given language determines the thought processes of the speaker of that language. I am not going to go into the detail of how the hypothesis was developed. The story is well-known. I believe it is sufficient to say that Whorf's idea was influenced by Sapir and that the chain goes back from Sapir to Boas and from Boas eventually to Humboldt.

Many linguists do not agree with Linguistic Relativity. Recently, researchers tested different versions of the hypothesis in the field of neuroscience, focusing on language and human communication. They ended up identifying a 'universal language network' in the brain. They affirm there is a genetically predetermined structural neural network. Despite huge differences in their languages, subjects of experiments demonstrated that key properties in their brains' language network were consistent with each other. That would not really falsify linguistic relativity.

Anthropologists and linguists in the field are constantly finding cultural differences that affect the way speakers of certain languages perceive the world, that is, that their thought processes are affected by their language.

Professor Shigeru Miyagawa *et al* conducted a study over the past eighteen years at MIT. The intention of the study was to determine when the first human populations had branched out. It was discovered that the first clear split between human groups had taken place some 135,000 years ago. Since every human culture has language, and languages have similarities, the inference was that the capacity for language pre-existed the separation. They believe the study confirmed that human language is monogenetic in origin and, thus, universal.

There are two problems with that conclusion: 1) the evolution of language did not happen overnight. It took tens of thousands of years, so it would be impossible to determine the exact period in which language had evolved to include complex syntax, which is a requirement for human communication; 2) the group from which the rest of the population branched out were the speakers of Khoisan languages; all Khoisan languages are 'click' languages, which means that some of their consonants are not common to most other languages. A clear taxonomy can be established between the two groups of languages: Khoisan and the rest. In fact, out of six thousand languages, approximately thirty have click consonants, most of them in South Africa. Apart from the ones in the original Khoisan languages, clicks may have spread from linguistic contact with the original group. There is only one exception that confirms the rule: Damin, which is a ritualistic language spoken by the Lardil people of Mornington Island, in Australia.

I previously mentioned another study conducted by Dr Andrey Vyshedskiy, from Boston University, concerning stages involved in language comprehension. Dr Vyshedskiy assures that 'time' is encoded in the parietal cortex, and that understanding of time evolved 70,000 years ago. Time is an integral concept of human cognition, which necessitates the existence of language. But there is a huge gap between that period and the one concluded by the MIT study for the existence of language (135,000 years). Could humanity have taken 65,000 years to comprehend time? It is unlikely.

But the time a language has taken to develop does not seem to affect the way in which the language-processing regions operate within the human brain. Another MIT study wanted to determine if 'natural' languages were processed differently from 'constructed' languages, like Esperanto or Klingon. The result was that these latter languages activated the same neural networks as 'natural' languages. The study apparently included some computer languages, which failed the test. None of this should have come as a surprise. Computer programming should not be considered a language. It is basically a series of mathematically expressed algorithms that have nothing to do with reality. Languages convey information and meaning about reality.

What was interesting in this study was that the human brain treated both 'natural' and 'constructed' languages the same. The reason is because they are both artificial. There are no 'natural' languages, in that language is not really biological.

Even the first language was devised by human beings. I suppose the definition of 'artificial' also comes into it. I consider that anything that is 'human-made' is artificial. Languages have resulted in cultures and influenced thought since the dawn of humanity.

ARISTOTLE, AVERROES & BORGES

- THE STORY OF A MISTAKE

The year was 1947. Jorge Luis Borges had moved with his mother, Doña Leonor, to a cozy two-bedroom flat on Maipú street, very close to Plaza San Martín, in the heart of Buenos Aires. The flat also had a bedroom for servants next to the kitchen, which was promptly occupied by the new maid, Fanny. The situation was not atypical or strange those days in Argentina. A forty-eight-year-old bachelor living with his mother and a servant in a small flat.

The months before then had not been easy for Borges. Estela Canto—his girlfriend—had unceremoniously dumped him. The newly-elected populist government of Juan Perón had terminated his employment in a suburban library. To add insult to injury, council bureaucrats had appointed him poultry and rabbit inspector, knowing very well that he was an up-and-coming literary figure in Argentina.

The months after the breakup with Estela, Borges had been writing his usual stuff. The months before, he had dedicated

"The Aleph"—his most famous story—to Estela. He had also written a few short stories that were later included in the homonymous book. The most important among those, by far, was the narrative that occupies us now: *"Averroes' Search"*.

The story is important for several reasons. Borges' writings are always deep, fascinating, surprising, often historical, sometimes philosophical. This one has several different readings and different levels. It may be taken as a quaint tale located in twelfth-century Cordoba, the cultural centre of *Al Andalus*. Some understand the story as supporting Linguistic Relativity—i.e., that language and culture influence the way we think. I concur. Analysing the narrative, the idea is well-founded and easily confirmed,. Averroes is a prisoner of his culture. But there are other important details worth mentioning.

IBN RUSHD, known as Averroes, was the most famous philosopher and commentator of the work of Aristotle in the Arab world; his writings—based in Spain as they were—reintroduced Aristotle to the West. The thought of Classical Greece had been forgotten during the Dark Ages, and Aristotle had become almost unknown then.

Based on a discovery by Renan of a mistake in one of Averroes' treatises, the *Tahafut-ul-Tahafut*, the story covers a few hours in the life of Averroes.

According to Borges, that day, Averroes is frustrated because two words keep popping up in Artistotle's *Poetics*: they are τραγωδία *(tragedy)* and κωμωδία *(comedy)*. Averroes cannot understand what Aristotle means by either of them.

Averroes looks out the window. Some boys are playing. They are mimicking the actions of adults—as children often do. One of them pretends to be the minaret, another one, the *muezzin*, and a third one represents the faithful. The *muezzin*, standing on the shoulders of the minaret, chants *"Allah il Allah"*. The faithful bows down with his head in the dust. The boys take turns as they repeat the game, but they all want to be the *muezzin*. Averroes—still thinking about the dilemma that occupies him—goes back to his books, ignoring their "calls to prayer". He has to attend a dinner at Farach's. Apart from being a friend, Farach is a Quranic scholar. One of the guests at the dinner is Abulcasim Al Ashari, a famous traveller, a kind of Arab Marco Polo. Abulcasim has been to many faraway places, including China.

During dinner, somebody comments on the beauty of roses. Abulcasim says he is convinced Andalusian roses are the best in the world. With a touch of humility, the host replies that —according to the wise Ibn Qutaiba—in Hindustan there are roses whose petals bear the inscription *"Allah il Allah, Muhammad Rassul Allah"* (*Allah is God and Mohammed is His Prophet*); surely, Abulcasim has seen them. In a difficult position, Abulcasim fails to answer. He mumbles that the Lord has the key to all things occult.

Averroes' authority saves the moment: it is easier to believe that the wise Ibn Qutaiba had made a mistake than admit there are roses with the Muslim profession of faith written on their petals. Somebody alleges there are trees whose fruits are supposed to be parrots. Averroes agrees that that would be more possible: both birds and trees are part of nature, whereas writing is an art. Another guest indignantly replies that the Quran, the 'Mother of All Books', predates Creation and is kept in Heaven. Averroes could reply but remains quiet.

Changing the subject, a guest asks Abulcasim to tell them about some marvel he has encountered during his travels. Abulcasim recalls something that happened while he was in Sin Kalan (Canton). Arab merchants had taken him to a house, which was actually a big room with many people eating and drinking in it. There were also people on a platform, some of them playing drums and a lute. Those on the platform would pray and sing and chat. They would be imprisoned, but the prison wasn't there. They would ride horses, but the horses were nowhere to be seen. They would fight and die, and then they would be alive again.

The dinner guests cannot understand how that could have been possible. Farach ventures that they were crazy.

Abulcasim replies that one of the merchants had assured him that they were not crazy, that they were showing a story.

Farach replies what appeared obvious to all those present: to tell a story you don't need many people. One speaker is enough.

The conversation then turns to poetry. Averroes discusses Arab lyric works and affirms with conviction that time enlarges the scope of any verse, and that the same happens with music. Then, he talks about the first poets, those of the Time of Ignorance, before Islam. All poetry belongs to them and to the Quran, he says. There is no place for innovation. The other guests are pleased.

Borges finishes the story the moment Averroes writes the mistaken definitions: *"Aristu (Aristotle) gives the name of tragedy to panegyrics and that of comedy to satires and anathemas"*. Suddenly, Averroes disappears. Borges has stopped thinking about him.

Borges then adds, as an afterword:

"In the foregoing story, I tried to narrate the process of a defeat... Later I reflected that it would be more poetic to tell the case of a man who sets himself a goal which is not forbidden to others, but it is to him. I remembered Averroes who, closed within the orb of Islam, could never know the meaning of the terms 'tragedy' and 'comedy'."

The summary I have attempted here may not be an apt replacement. The story is usual Borges—masterly written and worth reading in its entirety.

THERE ARE several points to the story. The first one and the most obvious is the one I have already mentioned: language influences the way the speaker thinks about reality. The story appears to confirm it. Averroes, immersed in a language and a culture, cannot understand terms that are alien to Islam and have no equivalent in Arabic. There were no words for *tragedy* and *comedy* in Arabic because the institutions were unknown in the Muslim world.

Averroes was a universalist; he developed a theory called "Unity of the Intellect", according to which all human beings share the same mind. With this story, Borges patently demonstrates that human reality has nothing to do with universals.

The work Averroes was working on was Aristotle's *Poetics*. We should remember that, according to Aristotle, the nature of poetry (which, at the time and in his view, included theatre) differs from that of philosophy in that poetry does not need an explanation, poetry mimics reality and includes emotions, while philosophy deals only with ideas: *"... there is no common term we could apply to Sophron and Xenarchus and the Socratic dialogues..."*. And when a philosopher writes using poetry, Aristotle maintains the distinction: *"...and yet Homer and Empedocles have nothing in common but the metre..."*.

Aristotle appears to divide consciousness along the same lines that divide sentience and cognition, i.e., a hybrid, layered, human consciousness. Poetry resorts to unusual meanings that add feeling to different terms. Drama has vision and sound, and it includes music. Drama appeals to the senses and explains how demonstration is easier to follow than explanation. Aristotle understands that both, poetry and drama, appear to be located where consciousness overlaps sensorial experience and intellectual understanding.

The way Aristotle deals with the concept of poetry is also illuminating in terms of the birth of drama. In the *Poetics*, playwrights and poets are not in separate categories. They are all poets of sorts. There appears to have been a progression. Epic poems must have been told and sung until gestures grew into drama.

THE STORY, however, has more related readings. Averroes failed to understand that the boys outside his balcony, through their game, were actually mimicking a story—they were showing something directly to the senses—and he missed his second opportunity at Farach's dinner party.

Children learn through imitation, or mimesis (μίμησις, in Greek). They imitate the behaviour of their elders. Mimesis is innate in all mammals. When children acquire language, however, the opposite process takes place, imitation is gradually replaced by explanation. A layer of cognition is superimposed on sentience. The process makes us an altricial species, one in which the young require prolonged rearing.

EXQUISITELY, Borges adds another marvel to the Muslim story. Somebody says there are roses that have calligraphy on their petals. Averroes denies that such roses may exist. His logic is impeccable: roses are natural, whereas writing is an artifice. Confronted with parrots that grow on trees, Averroes accepts that as a more plausible phenomenon. Parrots and trees are both part of nature. Writing is not. In the mind of Averroes there is a clear distinction between natural and artificial. Humans are used to the coexistence of sentience and cognition, but the miracle of a rose with writing on it beggars belief: it is impossible and anachronistic.

A FEW DECADES LATER, Umberto Eco, an avid reader and admirer of Borges, and someone who had evidently read *Averroes' Search* and the *Poetics*, would come up with *"Il Nome della Rosa"*, a medieval thriller that shows how Aristotle can be misinterpreted from a different, Christian, angle. This time, Jorge de Burgos, a blind abbot, hides the *Poetics* in a monastery's library, which is also a labyrinth (name and place, evident homages to Borges). The old fanatic monk doesn't want people to read the book because, as he argues, *"... qui si ribalta la funzione del riso, lo si eleva ad arte, gli si aprono le porte del mondo dei dotti, se ne fa oggetto di filosofia, di perfida teologia..."* (*... here the function of laughter is reversed, it is elevated to art, the doors of the world of the learned are opened to it, it becomes the object of philosophy, and of perfidious theology...*).

BASING his short story on a couple of anecdotes, Borges explains an error in Averroes' work by establishing that, during the times of the philosopher, theatre was totally unknown—and incomprehensible—in the Muslim world. Eco creates another misunderstanding of the work of Aristotle, this time from the angle of Christian fanaticism, and erroneously equating comedy solely with laughter.

It might be possible to add that poetry and drama use a limited amount of language in order to show something. Like the visual arts or music, they both demonstrate emotion. Any intellectual explanation of art is superfluous. Art and explanation have different natures. They appeal to different layers of human consciousness.

TIME

BORGES, AGAIN

- TIME AND OTHER IDEAS

"Your matter is time, the incessant time. You are every lonely instant."

- JLB

"Todo lenguaje es de índole sucesiva, no es hábil para razonar lo eterno, lo intemporal."

- JLB

*T*he way we perceive space is through vision, through the eyes. The way we measure it is also visual. What happens with time is something totally different. Although time-keeping can be visual, the perception we have of time is exclusively cognitive. In the second quotation above, Borges discusses language and eternity. Language is incapable of dealing with eternity, with timelessness because it includes time in itself (obviously Borges is not considering Hopi here). We know that time exists, at least for humans, because it appears to be part episodic memory (the past), and part expectation (the future).

The only thing that happens in reality is change. Time, which Aristotle defined as *"the measure of change"*, is definitely a human construct. Animals live in a constant present, they have no time, no expectations, no regrets. Some species appear to have some degree of memory, but no long-term episodic memory. Of course, they cannot strategise and they do not fear death as something inevitable. Finitude also resides within cognition.

We can see how things change, how a creature grows, how a tree grows, how a flower withers or how we age, but those changes are almost imperceptible to the senses. Temporality lives within us, in the cognitive layer of our consciousness.

IMPORTANT INNOVATIONS usually include considering ideas outside pre-established frameworks or canons, thinking divergently. We already saw how Newton devoted his earlier years to alchemy, like many others of his time, until he came up with something that no one had thought of until then: he began to apply what we now know as a scientific discipline to his discoveries. So did Descartes with his *Discourse on Method*.

Many Western thinkers agreed with their counterparts in the East and could see that time was just a human illusion that resides within cognition. In 1818, Arthur Schopenhauer could understand that the present is the only reality: *"Above all things, we must distinctly recognise that the form of the phenomenon of will, the form of life or reality, is really only the present, not the future, nor the past. The latter are only in the conception, <u>exist only in the connection of knowledge, so far as it follows the principle of sufficient reason</u> *. No man has ever lived in the past, and none will live in the future; the present alone is the form of all life, and is its sure possession which can never be taken from it. The present always exists, together with its content. Both remain fixed without wavering, like the rainbow on the waterfall.". (The World as Will and Idea).** [my underlining].

TO READ BORGES IS, among other things, to observe how a lucid mind faces two important, but more than anything, universal issues: consciousness and time. We know that Borges has other obsessions and that he constantly plays with them. He is obsessed with tigers, mirrors and labyrinths. However, all his writings reflect, sometimes tangentially, his two fundamental themes which—I repeat—are consciousness and time. And he does it from the perspective of an individual who is Argentine and Western, but who never finishes accepting those limitations and embraces his humanity, like Hesse and Schrödinger, even venturing into Eastern ideas and experimenting with a mindset that ignores the limits imposed by Aristotle.

Sometimes Borges denies objective reality. He repeats again and again that one man is all men and that to kill one is to kill humanity. Consciousness is one and we share it in space and time. *"In short, immortality exists in the memory of others and in the work we leave,"* he tells us.

IN *THE IMMORTAL*, Borges creates the character of Joseph Cartaphilus (Latin for *'Lover of paper'*), someone who is immortal and remembers being, at the same time, Marcus Flaminius Rufus, a Roman centurion, and Homer. Cartaphilus is three people who are, in reality, one. He is not eternal, he is immortal. At one point the character tells us: *"Being immortal is trivial; except man, all creatures are immortal, for they are ignorant of their death; what is divine, terrible, incomprehensible, is to know oneself immortal."* Here, no doubt, he is telling us something about the duality of consciousness. Finitude comes with cognition. Cognitively, immortality is incomprehensible.

In almost all his essays and stories, from *History of Eternity*, to *Funes, the Memorious*, to *Garden of the Forking Paths*, or *The Script of God*, Borges deals with the existence or non-existence of time and all its possibilities.

History of Eternity shows us the religious side of someone who claims to be agnostic: *"The universe requires eternity.*

Theologians are not unaware that if the Lord's attention were diverted for a single second from my right hand that writes, it would fall into nothingness, as if it were struck by a fire without light. That is why they affirm that the conservation of this world is a perpetual creation and that the verbs 'conserve' and 'create', which are hostile to each other here, are synonymous in Heaven.". Like Schrödinger, Borges' views on life and entropy are clear.

If we want to learn what his ideas about time are, all we have to do is read his *New Refutation of Time*. In that essay, Borges begins by making us see that the title contradicts what he says in the essay: that the continuity of time is an illusion. Time does not contain a succession. Every moment is eternity, which negates the mere inclusion of the word "New" in the title. Here, as we will see, Borges agrees with Rovelli, the physicist, and the experts in quantum mechanics. In other writings, he often mentions eternity, for example, when he says, *"[Eternity] ... theologians defined it as the simultaneous and lucid possession of all past and future moments, and judged it one of the attributes of God."* Other times he references things like the duration of hell, but those opportunities are almost always an homage to Dante, or perhaps Swedenborg.

With his reasoning, icy and very clear as always, Borges tells us about his search, mixing the idealism of Berkeley with the ideas of Hume. The latter rejects identity. He says that each man is a collection of perceptions that occur one after another with inconceivable rapidity. Both believe in the existence of time: for Berkeley it is a succession of ideas. Hume says it is a sequence of indivisible moments. Borges flirts with the two perspectives, takes sides decisively and, in doing so, proposes something new: *"I have accumulated transcriptions of the apologists of idealism, I have lavished their canonical passages, I have been iterative and explicit, I have censured Schopenhauer (not without ingratitude), so that my reader would enter that <u>unstable mental world. A world of evanescent impressions</u>; a world without matter or spirit, neither objective nor subjective; a world without the ideal architecture of space; a world made of time, of the absolute uniform time of the Principia; an indefatigable labyrinth, a chaos, a dream. To that almost perfect disintegration came David Hume."*

Borges' inimitable erudition brings us back to the idea of the ego, which Eastern philosophy and Buddhism found illusory thousands of years ago. That, he tells us, only rejects the notion of the time we imagine knowing. He tells us about Zeno's paradoxes, which oppose pluralism and change, and claim that movement is nothing but an illusion. There are truths that seem to deny what appear evident to our senses.

Berkeley, on the other hand, (*Principles of Human Knowledge*), rejects the primary qualities—the solidity and extent of things—and absolute space. Borges suggests that if we reject the existence of matter and spirit, and deny space as well, we have no right to retain time as continuity. Time, then, does not exist outside of the now. But, in that now, time is everything.

Once he admits idealism, Borges goes much further: he explains the dynamic nature of our identity, telling us "... *there is no secret self behind faces, which governs acts and receives impressions; we are only the series of those imaginary acts and those wandering impressions.*" This is the acceptance of pure sentience. The moment we deny the ego and accept only change, we are stepping outside the Western idea of cognition and objective reality.

The *New Refutation* continues with a disquisition on collective consciousness and the unity of humanity: *"... if there is no plurality, he who annihilated all men would be no more guilty than the primitive and solitary Cain, an orthodox view, nor more universal in destruction, which can be magical."* There is not a multitude of sorrows, there is only one pain. He who kills one man, kills all. The perception of what is real is something we share without being objective. We all perceive everything. Furthermore, he is saying that, in reality, we are iterations of *H. Sapiens*.

In his writings, Borges expands on granularity, the same as the one in quantum mechanics: according to Anaxagoras, he tells us, gold consists of gold particles, and according to Josiah Royce, everything present is a succession, and he tells us that our language is not suitable to explain timelessness or eternity. The unsuitability of language to express certain things is pure Wittgenstein. Time, however, exists within cognition. What is ineffable is timelessness or eternity.

Trying to describe the *New Refutation...* analysing the essay is an exercise in futility. You have to stick to the work. Recommended reading. It's pure luxury.

Borges—who defines his work as *"the weak artifice of an Argentine lost in metaphysics"*—can only conclude all his speculation with a mixture of truth and poetry: *"And yet, and yet... Denying temporal succession, denying the self, denying the astronomical universe, are apparent despairs and secret consolations. Our destiny (as contrasted with Swedenborg's hell and the hell of Tibetan mythology) is not frightful by being unreal; it is frightful because it is irreversible and iron-clad. Time is the substance I am made of. Time is a river which sweeps me along, but I am the river; it is a tiger which destroys me, but I am the tiger; it is a fire which consumes me, but I am the fire. The 'world, unfortunately, is real; I, unfortunately, am Borges."*

The author, who does not discuss directly the possibility of a two-layered consciousness in his writings, nor ever mentions time as a cognition-related phenomenon, gives us this clear hint that he would have agreed with those criteria: *"But we do not even possess the certainty of our poverty, inasmuch as time, easily denied by the senses, is not so easily denied by the intellect, from whose essence the concept of succession seems inseparable".*

SCHRÖDINGER, AGAIN

- TIME ACCORDING TO SCIENTISTS

"If time is a mental process, how can it be shared by thousands of men, or even two different men?" ... *"None of the various eternities that men planned—that of nominalism, that of Ireneaeus, that of Plato—is a mechanical aggregation of the past, present, and future. It is a simpler and more magical thing: it is the simultaneity of those times. The past is in its present, as is the future. Nothing happens in that world, in which all things persist, still in the happiness of their condition."*

A new refutation of time ——

Jorge Luis Borges

The laws of physics are clear, if something can be measured, quantified, defined mathematically, is an observable quantity on which other observable variables depend, that something exists.

To be real, something has to meet all those criteria. In physics, if something is not possible, it is called a "pathology". Maybe time is pathological? Maybe it's impossible? Well, it meets all the conditions above to be possible. It would have to be real. The problem is that the answers to all those questions are relative. Einstein proved that time is real, although it does not seem to be real in an objective way. How so? According to the theory of relativity, time is not pathological, it is only relative. So, the idea that time is relative doesn't prove that time doesn't exist. Change exists as a subjective experience and time, which measures it, is produced and communicated cognitively. It is only human. So, change is felt intuitively, subjectively. When we keep time, figures give us an 'objective' approximation of how much change there has been.

There are differences between Einstein's theory of relativity and the more recent discoveries of physics regarding the quantum properties of space and time. What appears to be the consensus now is that the temporal part of the theory of relativity disappears the moment we consider the quantum perspective, that is, from the moment we consider the world at a minute level.

What has quantum mechanics discovered with respect to time? Well, three fundamental characteristics: its granularity, its indeterminacy, and its relationship with other physical variables. Problem is that there is a scale, called the "Planck scale", that measures the tiniest chance of time in the gravitational field. The smallest chance of measuring time is called

"Planck time". It is $[5.319124 \text{ x}]^{10.44}$ of a second. That is, a hundred millionths of a trillion, a trillion of a trillion of a second. Planck's time, then, cannot be measured. No current clock can. As mentioned above, according to physics, if it can't be measured, you lose a condition for the existence of something. Then, time does not exist. Well, apparently, it does to a certain extent.

Carlo Rovelli, the eminent Italian physicist, writes in his book *"L'ordine del tempo"* that granularity is a universal characteristic: *"Perhaps the rivers of ink that have been spent talking about the nature of the 'continuous' through the centuries, from Aristotle to Heidegger, have been wasted. Continuity is just a mathematical technique to approximate things of very fine granularity. The world is subtly discrete, not continuous. The good Lord did not design the world with continuous lines: he did it with a light hand, he sketched it with dots, like a painting by Georges Seurat."*. Although we may not agree, it must be said that the man explains it in a brilliant way.

How did we come up with the concept of time? Was there a beginning to time-keeping?

We know that many changes happened to humanity before we were born. Some were recorded others weren't. Some changes were lost to history. Some were remembered orally for generations. But, —and this is very important— in order to explain reality with all the changes that happened before us; in order to explain our long-term memory, we *invented* time.

When we say "time immemorial" we are talking about a time that existed before our collective memory. We have no evidence that it existed. In fact, there was no such time. All the records we have are archaeological. Before human consciousness, our ancestors were just like all other animals. There was change, individuals lived, died and were born, but without individual or group identity, nobody kept track of when those events happened; nobody knew who did what. Time was not measured because nobody knew anything except change. Information and communication were scarce. History, which is recorded change, did not exist.

The beginning of time is simultaneous with the beginning of long-term memory. All there was before memory was change. When we had not developed culture, when we were like all other animals, in effect, we had no time.

According to recent studies, the first—plausibly mnemonic—notations about seasons and parturition of prey were recorded in caves throughout Europe some 40,000 years ago. But that was 'a recorded' notion of time. Before then, we are told that time became encoded in the parietal cortex. No other species appears to have time or understand the concept of time.

The phenomenon of European cave paintings is an interesting one. Until very recently, they were considered art. The information we have now places those paintings well outside the sphere of art. The figures are icons of animals: all of them are prey. Together with each icon, there are a few dots or lines and a "Y" towards the end. What palaeoarcheologists surmise is that those notations carry a message: parturition of this animal (icon) occurs after so many lunar months from the main hunting season, or something similar. Without a non-iconic way to record the type of prey, the best way was to draw the animal. The dots would have been a very primitive numbering system, and the "Y" meant birth.

The difference is significant. We discuss art elsewhere in this book, and it is totally about perception and feeling. A record is about sharing information. In this case the information is time.

BUT LET us see what Schrödinger said about time. One of the most important contributions of science—according to the physicist—was the *"gradual idealisation of time"*. Of course, that had happened long before science became science as a separate discipline. He refers to how humans passed from measuring cycles, to time-keeping, to an ideal concept of time. How time became part of human knowledge (or, to put it in other words, how we invented time). Schrödinger mentions Plato as the first to contemplate the possibility of a "timeless existence". He then discusses causation: *"Time is the notion of 'before and after'. ... The notion of 'before and after' resides on the 'cause and effect' relation. We know, or at least we have formed the idea, that one event A can cause, or at least modify, another event B, so that if A were not, then B were not, at least not in this modified form."*.

Schopenhauer uses different words to say the same thing: *"This simplest form of the principle [of sufficient reason] we have found to be time. ... The past and the future ... are empty as a dream, and the present is only the indivisible and unenduring boundary between them."*.

The effect cannot precede the cause. That is fine, but then Schrödinger admits that causation—which is clear in mathematical language—becomes muddled because *"everyday language is prejudicial in that it is so thoroughly imbued with the notion of time — you cannot use a verb (verbum, 'the' word, Germ. Zeitwort) without using it in one or the other tense."*. The fact is that there are events that are neither earlier nor later than A. *"The region of space-time occupied by this class is called the region of 'potential simultaneity'."*. Potential simultaneity becomes the liberation from causation.

Within sentience, time exists solely as change, but then human cognition explains it through causation and then measures it, and then it really becomes time. Time-keeping is not an explanation of change, it only measures it.

Schrödinger then discusses the unidirectionality of time. Time goes from past to future, not the other way around. That is linked to the Second Law of Thermodynamics: the entropy of isolated systems left to spontaneous evolution cannot decrease, as isolated systems tend to thermodynamic equilibrium, i.e., they rest. Without interference, whatever is hot always becomes colder, not the other way around (mechanical or statistical theory of heat). According to Schrödinger, that theory *"has an even stronger bearing on the philosophy of time than the theory of relativity. The latter, however revolutionary, leaves untouched the unidirectional flow of time."*. In any case, he concludes that—because time is a creation of cognition—physical theory suggests that mind cannot be destroyed by time. That goes without saying. Time is a human construct. It's the other way around. Without consciousness time will cease to exist. It did not exist before consciousness.

PHILOSOPHY

WITTGENSTEIN

- UNDERSTOOD BY FEW

"What is your aim in philosophy? - To show the fly the way out of the fly-bottle."

Ludwig Wittgenstein

In a previous chapter I state that Gautama had evidently guessed the dual nature of human consciousness. Not many other people in history did. Ludwig Wittgenstein was probably one of them. I do not know to what extent Wittgenstein saw cognition as an artificial, meta-evolutionary, addition to human consciousness. He did not state it in so many words, but his writings, especially his *Tractatus Logico-Philosophicus*, indicate a profound understanding of the issue. The counterpart is that many philosophers—notably his teacher and mentor, Bertrand

Russell—guessed they were before a brilliant intellect, but failed miserably to understand his message.

That Wittgenstein's theories were incomprehensible to many is evidenced by the fact that now, almost a century later, science and philosophy are still trying to understand, explain and measure human experience.

WITTGENSTEIN WAS the scion of a very wealthy Austrian family. His father, Karl Wittgenstein, was one of the most powerful industrialists of his time. Ludwig had four older brothers and two sisters who were also bright in different ways, but by the time he had reached his twenties, it was evident that he was a uniquely gifted individual.

Following in his father's steps, he started studying mechanical engineering in Berlin. By 1908, he was training in aeronautics at Manchester University. While conducting his research in aerodynamics he invented a special propeller. In order to solve the problems of his design, he studied the mathematics involved. One of the pioneers in mathematical logic at the time was Bertrand Russell, so Wittgenstein applied to study under Russell's guidance. Within two years, Wittgenstein had nothing to learn in that field and was arguing with Russell about his theories. He soon abandoned engineering and devoted all his energies to the study of philosophy. He discovered that all of the great philosophers had made "disgusting mistakes".

Wittgenstein returned to Vienna and, at the outbreak of WWI, volunteered as a private in the Austrian Army. Eventually he was given a commission and sent to the Italian front. Towards the end of the war, he was taken prisoner by the Italians and confined in Monte Cassino. During the intervening years he had been writing the *Tractatus Logico-Philosophicus*, the manuscript of which he kept with him while he was a prisoner of war.

WITTGENSTEIN WAS NOT interested in debates. He expected to be clearly and completely understood. Otherwise, there would be no point in saying anything at all. Bertrand Russell once told him that, rather than just stating what he thought was true, he should provide arguments. Wittgenstein replied that providing arguments would spoil the beauty of the idea. The *Tractatus* is based on that kind of logic. It is either clear, or it is unclear. We believe it has been mostly the second.

FROM RUSSELL'S *Introduction* to the *Tractatus*, it becomes apparent that he had not quite understood what Wittgenstein was saying. The book is important—he says—and it is not incorrect, but he cannot explain why: *"I find myself unable to be sure of the rightness of a theory, merely on the ground that I cannot see any point on which it is wrong. But to have constructed a theory of logic which is not at any point obviously wrong is to have achieved a work of extraordinary difficulty and importance"*. Knowing Wittgenstein personally, understanding his genius, but unable—or unwilling—to say that he had not understood, Russell stated: *"There are*

some respects, in which, as it seems to me, Mr Wittgenstein's theory stands in need of greater technical development". [Please, Ludwig, explain what you're saying!!]

Russell was revered as one of the most important logicians and thinkers of his time (and to some extent, he still is). To admit that he had not understood what's probably the ultimate theory within his discipline was courageous and pathetic at the same time. He was the first one no to understand him, but he would not be the last one. More than a century has passed and it is quite obvious that Wittgenstein remains misunderstood and ignored by the majority of philosophers, logicians and scientists.

WITTGENSTEIN WAS QUITE clear in that he didn't care. He wrote for his peers: *"This book will perhaps only be understood by those who have themselves already thought the thoughts which are expressed in it—or similar thoughts."*

He famously stated that trying to explain sentience was an exercise in futility. There is no other way of understanding him *"[This book's] whole meaning could be summed up somewhat as follows: What can be said at all can be said clearly; and whereof we cannot speak thereof one must be silent."*

The philosopher clearly states that the idea of time is a human construct: *"6.3611 - We cannot compare any process with the 'passage of time'—there is no such thing—but only with another process (say with the movement of a chronometer). Hence the description of the temporal sequence of events is only possible if we support ourselves on another process."* In this case, the movement of a chronometer is a product of cognition. Time can only be understood in human terms. There is no other way.

But he goes even further and rejects causation as well: *"6.36311 - That the sun will rise to-morrow is an hypothesis; and that means that we do not 'know' whether it will rise. 6.37 - A necessity for one thing to happen because another has happened does not exist. There is only 'logical' necessity. 6.371 - At the basis of the whole modern view of the world lies the illusion that the so-called laws of nature are the explanations of natural phenomena."*. Evidently "laws of nature" is a misnomer. Nature has no laws. Scientists have invented principles that explain certain phenomena or behaviours of nature to a human mind.

Without even mentioning the dual nature of human consciousness, Wittgenstein constantly reflects on the impossibility of an answer: *"6.52 - We feel that even if 'all possible' scientific questions be answered, the problems of life have still not been touched at all. Of course there is no question left, and just this is the answer."*. Nature needs no explanation. It just is. Human beings— after developing cognition— need to understand, hence, religion, philosophy and science.

The paradox is that in order to understand we have to explain that there is something inexplicable: *"6.54 - My propositions are elucidatory in this way: he who understands me finally recognizes them as senseless, when he has climbed out through them, on them, over them. (He must so to speak throw away the ladder, after he has climbed up on it.)"*.

We could go on and on quoting Wittgenstein from the *Tractatus* or from other sources. Yes, there are some inconsistencies here and there, but the only conclusion we can reach after having read him is that what he wanted to state was totally ineffable. He did it, though. He discussed human consciousness, culture, time, and different views of the world, and then he did what someone like him could only do: he told us to throw away the ladder and stand on thin air. That is what I am trying to do here.

SCHOPENHAUER

- THE ORIGINAL WESTERN VISIONARY

We have said before that a few Western thinkers had been impacted by the profoundness of Eastern thought, and we have given examples. What can we say about Schopenhauer? We can say, for instance, that he was the first important Western philosopher to acknowledge the value of the Upanishads. Although he was deeply influenced by Immanuel Kant's idealism, Schopenhauer's thought and his connection with Eastern mysticism influenced, in turn, thinkers like Wittgenstein, and others, who recognised the important difference between sentience and cognition in human consciousness.

Schopenhauer used his own terminology, which sometimes makes his thinking difficult to follow in important works like *The World as Will and Idea,* but that minor obstacle disappears once you become accustomed to the terminology. The importance, depth and width of his contribution to Western philosophy cannot be underestimated.

The message is quite clear: he bases his philosophy on Kant, and recognises idealism in general, but there is a strong Eastern feeling in much of his thinking, as he acknowledges early in his work: *"The philosophy of Kant, then, is the only philosophy which a thorough acquaintance is directly presupposed in what I have to say here. But if, besides this, the reader has lingered in the school of the divine Plato, he will be so much the better prepared to hear me, and susceptible to what I say. And if, indeed, in addition to this is a partaker of the benefit conferred by the Vedas, the access to which, opened to us through the Upanishads, is in my eyes the greatest advantage which this still young century enjoys over previous ones..."*. But Schopenhauer was preoccupied with sentience, with how we have perception. Kant left perception as a fundamental, something that comes from the outside.

The Western side of Schopenhauer's idealism is nothing new: *"All that in any way belongs or can belong to the world is inevitably thus conditioned through the subject ... The world is idea. ... [This] was implicitly involved in the sceptical reflections from which Descartes started. Berkeley, however, was the first who distinctly enunciated it..."*. But then he points to the early recognition of this truth as a fundamental tenet of Vedânta philosophy. Schopenhauer reflects on both layers of human consciousness: *"These words adequately express the compatibility of empirical reality and transcendental ideality."*.

The coexistence of experience and thought within human consciousness was accepted until the second half of the twentieth century, when behaviourists argued that, being unable to conduct scientific studies on sentience, they had to sideline cognition.

∽

An inveterate critic of European academia's *"cherished mediocrity"*, Arthur Schopenhauer was born in 1788, the son of a well-to-do German family; he attended Gottingen and Berlin Universities, where he studied medicine, philosophy, metaphysics, logic and psychology. His dissertation was *"On the Fourfold Root of the Principle of Sufficient Reason"*. He travelled widely, visited Italy, where he lived for one year, and spoke Italian fluently. He could also speak Spanish, Greek and several other languages.

∽

Schopenhauer stresses the difference between *"ideas of perception and abstract ideas"*. He understands that that difference is among the most important facts regarding human consciousness. *"... another faculty of knowledge has appeared in man alone of all earthly creatures, an entirely new consciousness, which, with very appropriate and significant exactness is called 'reflection'. For it is in fact derived from the knowledge of perception, and is a reflected appearance of it. But it has assumed a nature fundamentally different."*. He saw with incredible clarity the causality between language and cognition: *"Speech is the first production, and also the necessary organ of his reason"*.

The connection with Eastern traditional thought is everywhere in Schopenhauer's work. It is totally pervasive, a constant; together with Gautama Buddha, he identifies cognition—which he calls *'willing'*—as the source of all human stress and sorrow: *"All 'willing' arises from want, therefore from deficiency, and therefore from suffering. The satisfaction of a wish ends with it; yet for one wish that is satisfied there remain at least ten which are denied."*.

AFTERWORD

My conjectures are summed up here:

Originally, humans were born, like all other animals, biologically equipped with sentience without a division—I would guess—between the internal perception of the individual's own body and the perception of the rest of nature. An iteration of the species, the individual was one with all of reality. There was hurt, there was pain, and affection, and instinct. Danger was perceived. Good food and sexual encounters were enjoyed but, in general, there was no analysis of reality. No divisions between 'I' and 'otherness'.

After hundreds of thousands of years as one animal species among many, our hominin ancestors developed—created and developed would be more appropriate—complex language. Renowned linguist Daniel Everett—the one who produced the Pirahã grammar—talks about the 'invention' of language. In any case, it took a long time.

Complex thought developed together with recursive language. It was a revolution. The way I see it, it was more than that: it was a meta-evolutionary event of unique proportions, only comparable to the origin of life on Earth. Humanity was established by adding that artificial layer of consciousness. Pre- and post-language layers co-exist to this day in our consciousness—that, I believe is quite evident. Once you understand the history, understanding consciousness becomes much easier.

Cognition seeks explanations, even for its own environment (human consciousness). Thought, when expressed, analysed and verified—or falsified—becomes knowledge. Knowledge is the aim of philosophy and science.

Sentience, the original layer, is the natural one. Ineffable, it requires no explanation for anything, basically because explanations are cognitive—a human artifice—and sentience is purely biological. We share it with our fellow animals, who need no explanations for their lives, behaviours or perceptions.

Western religion, philosophy and science provide different answers. Some may be true, some may be fiction, but they all are and will be limited to what is explainable.

Having failed to understand Wittgenstein and Schrödinger, but especially Wittgenstein, scientists and philosophers obstinately try to explain experience—what is worse, they fail to understand that, perforce, human consciousness has to include a discrete component: cognition. Paradoxically, trying to explain or measure sentience does not make any 'sense' either.

AFTERWORD

The philosopher is also misunderstood by religious believers —like Michael Egnor—who do not appear to fully comprehend the discreteness and different nature of sentience:

"[Wittgenstein says] that there are things of such fundamental importance to us that efforts to speak of them inherently mislead us. I believe that consciousness is such a thing, and that is why it cannot be defined. We cannot say clearly what we mean by consciousness, not because we haven't gotten the philosophy right or because we need more neuroscience experiments, but because conscious experience is too close to us, too fundamental to us to put in words. I think that Koch and, hopefully, other neuroscientists and philosophers are coming to understand this.

We cannot define consciousness or explain it by the methods of logic or neuroscience. Consciousness is that by which we perceive and understand, not that which can be understood."

Yes, but the notions are jumbled up: what cannot be understood is only sentience. The moment he uses the word *"understand"* he is referring to cognition. And cognition can be understood.

The issues and the solutions posed by the brilliant minds I mention and cite in this book—and the conjectures I submit—are transcendental. Some of the conjectures may be highly controversial and subject to criticism, maybe even ridicule. There are many more unanswered questions. Let us deal with some of them.

AFTERWORD

What does it mean that we are an "altricial" species? The term altricial comes from Latin *alere*, which is to nurse, to rear, to nourish. It is the opposite of precocial. The young of some species are born well-developed and mobile. Giraffes can run with the herd the same day they are born. Chimpanzees take about three years to be weaned; they become adults at a very young age. A human individual takes about twenty years to become a fully operational adult. On average, humans live to seventy years of age. That means that growing up takes almost one third of our lives.

Human individuals are born utterly inadequate. Why is it so? Well, to survive, we need to live in human society, our culture, which is as artificial as language and thought. We have to be reared biologically, and then cognitively, that is, we learn to speak, then we are socialised, and then schooled. Learning to speak means that we have to learn a particular language. The words we utter and understand are something artificial created by a particular culture. If not, all human beings would be speaking the same language from the time we were born. When we learn to think, we do so, aided by that particular language, which—I repeat—is not universal. Literacy is an important part of our schooling. We learn to read and write. Those are secondary skills that come after listening and speaking, which are the more basic linguistic skills. Then we learn something even more abstract: numeracy. And then, a mixture of both, algebra.

Failure to socialise and educate a human child results in the incomplete development of the individual. The human being reverts to its solely biological condition. This has happened on several occasions. The absence of human role-modelling can have amazingly sad results. Through history, there have been several cases of children who grew up surrounded by animals and learned their behaviours. One of the most recent ones is that of a Ukrainian woman, Oxana Malaya, who was born in Kherson Oblast, of alcoholic parents. When she was three, her parents left her outside to fend for herself. The toddler looked for warmth and protection among the dogs that were around the house. She was literally raised by those dogs. She lived in a kennel. All her contact with other beings occurred there. She became one of the pack. By the time a government agency rescued her, she no longer behaved like a human being. She walked on all fours, and could not speak. She barked and growled and, in general, her behaviour was that of a dog. After rehabilitation, Oxana has a limited vocabulary and her mental capacity is that of a five or six-year-old. Unfortunately, she lost the opportunity to learn how to behave like a human being at a time when she needed role-modelling. We definitely need to be socialised when we are toddlers. In order to reach our potential and become fully human, we must be brought up by human beings.

Other cases are historical, pseudo-historical or fictional: one of them is an experiment apparently carried out by Holy Roman Emperor Frederick II, who, in the thirteenth century, ordered that a few babies be brought up without any language input or physical contact. The idea was that they would be speaking the universal language that God had given Adam and Eve (he could not decide whether it was going to be Sanskrit, Hebrew or Latin). Eventually, the babies grew to become toddlers unable to speak, or died because of lack of human contact. Other experiments reported were conducted by Akbar the Great, a Mughal emperor, in the sixteenth century, and James IV of Scotland, in the fifteenth century. Of course, all results, real or fictional, were negative. Children cannot learn to communicate without input from the parents and/or the collective.

If we accept that, then we accept that language is not something natural. It does not come with us as part of our biological equipment. Evidence of it is that we have to learn it every generation, and that there is no universal language.

We never question that because human beings have been partly artificial from the beginning of humanity. That happened tens of thousands of years ago. *H. sapiens [the man who knows]*, is exactly that, the first hominin who can think, and we can think mainly because of language. When our ancestors 'invented' language—as Everett would say—meaning came with it. We have mentioned before that the process that took us from grunts and yells to meaning and words and syntax took many thousands of years. In any case, the origin of our thinking skills is foggy but unquestionably true. One day we grunted, the next, we were speaking and we had become human. Given that it was such a long period, determining the exact moment when an evolutionary process became meta-evolutionary and we could communicate complex thoughts would be impossible. We are sure, however, that one day our species transitioned from non-cognitive animals to cognitive human beings.

There are other species that are partly artificial as well—like cats and dogs, poultry and cattle. They are not totally natural because we have bred secondary species derived from their original natural species. Cattle were originally wild auruchs; dogs were wolves, and so on, until we decided to alter their genetic material to suit our purposes. They are partly artificial because they are dependent upon our species; and we use them for transportation, food, etcetera.

In the case of our species, our distant ancestors pulled themselves up by their [non-existent] bootstraps and, from the animals they were, they became fully human. Going totally against nature (against normal evolution), they accomplished something that even today appears impossible.

A recent study has discovered a protein that—according to those conducting the study—may have sparked the origin of human language'. That conclusion purports to explain that the ancestors of *H. sapiens* acquired language while *Neanderthals* and *Denisovans* did not because our species had a special, species-specific variant of the NOVA1 gene, called I197V. Why would we have that gene and the other *Homo* species not? The answer is that, although NOVA1 may have contributed to the evolution of human language, we did not have NOVA1 before 'inventing' language, if you like. It's the other way around: the gene came into being as a result of behaviour. Cells turn on or suppress certain genes as a result of their environment, in this case, the behaviour of our ancestors. Gene activation definitely follows environmental changes. I would argue that language—and cognition, in a feedback loop—were part of a meta-evolutionary phenomenon that resulted in the origin of our species.

∼

We have been preoccupied by the mysterious origin and nature of our consciousness for quite some time. How come we are the way we are? How are we different from other animals? Well, apart from the fact that we are alive like them, many things make us different from them, if we want truisms: we wear clothes, we can communicate with each other, we can keep time, we are creative, we are artistic and adventurous, our cultures decide what is right and what is wrong, we have identities and free will, we understand abstract concepts, etc. We have divided ourselves into ethnic groups and geographic areas with which we identify, and which we call countries, we have governments of different types, religions, we trade, etc.

Other animal species are creative and adventurous, one might say. Yes, but the big difference lies in the nature of their creativeness, and their adventurousness. The nature of their skills is almost totally instinctive. How can you tell the difference? Well, beavers build dams, but that is all they can build. They cannot build anything else. Of course, they wouldn't be able to build a tennis court. That would be too complicated. But you cannot ask a beaver to build a bridge with the same logs he uses for the dam either. He builds dams instinctively. The same thing happens with a bird's nest or a spider web. A swallow and other migratory birds travel long distances; a whale travels long distances. But they are not adventurous. They always follow the same routes for specific purposes. That is because their travelling is instinctive. Humans can choose where they want to go, or decide on different routes taking dangers or obstacles into consideration.

There is no doubt that animals have some degree of rationality. Crows solve problems and use tools to obtain food. That means some cognition. So do chimpanzees and other animal species. What they do not have is complex thought, recursive language or metacognition. They cannot speculate about their own thoughts, neither can they convey complex thinking to other individuals because they lack language.

I would say that neither Hume nor Descartes were right about the reasoning of animals. Descartes compared them to automata: no cognition at all. Hume said they were *"endowed with reason and thought"* like human beings: full cognition. The way I see it, the truth lies somewhere in between those two extremes. Animals cannot carry out complex tasks, neither can they communicate complex ideas because they cannot think conceptually, or abstractly, if you prefer. As a linguist I would tend to say that human beings can think complex thoughts, solve abstract problems and co-operate on large enterprises because we have recursive language. Language is a unique phenomenon that has given us cognition beyond the limited degree other animal species have. What animals have is qualitatively and quantitatively negligible compared to human cognition. Perhaps it should not be treated as full reasoning. I would argue that causal reasoning (the ability to make a tool in order to obtain something), based on the input of the senses, cannot be compared to how human beings understand the relations that may exist between thoughts. That really is cognition. Purely human cognition.

I submit here that AI will never reach the level of human consciousness. AI cannot be sentient by definition. Its nature, as the name indicates, is totally artificial. Sentience is biological. Furthermore, large language models like Chat-

GPT, do not actually 'understand' language. ChatGPT is a logarithm with a massive memory. Operators input millions of sentences and the logarithm statistically learns the exact words to use within a certain context. It's a Bayesian process. That does not mean it thinks or understands. The logarithm can also fake being sentient, but that comes a result of input. It cannot begin to do anything without outside input because it has no agency.

In order for an automaton to achieve something similar to human consciousness, it would need to be hybrid, like us. Human consciousness requires cognition, but it also requires the live biological component.

In this book I attempt to prove that the components of our consciousness—its layers—have different natures. Why do I say that? When we analyse why, we find that sentience—originally the sole component of our consciousness, like those of other animal species—is natural. Cognition, however, is the artificial creation of our human ancestors, the result of a "black swan" process of meta-evolution. From our perspective, human consciousness appears to be a hybrid phenomenon.

The irony of it all is that we have always been what we have long feared we would become: artificial, or at least partly artificial. Human beings are not natural animals. We have an extremely important artificial component, which is our cognition.

ACKNOWLEDGMENTS

Inés, of course. As usual, she read all the versions, criticised and helped edit all of it.

John Watts read some chapters, criticised and offered alternatives.

Rod Haedo read much of it and helped with ideas.

My brother Patocho listened to ideas, criticised and offered alternatives.

Andrey Vyshedskiy helped with inspiration, replies to my emails, and confirmed conjectures with his comprehensive studies.

My debt of gratitude to all of them.

www.ingramcontent.com/pod-product-compliance
Lightning Source LLC
Chambersburg PA
CBHW071849070526
44583CB00016B/1608